Knowledges in Publics

Knowledges in Publics

Edited by

Lorraine Locke and Simon Locke

CAMBRIDGE
SCHOLARS
PUBLISHING

Knowledges in Publics, Edited by Lorraine Locke and Simon Locke

This book first published 2013

Cambridge Scholars Publishing

12 Back Chapman Street, Newcastle upon Tyne, NE6 2XX, UK

British Library Cataloguing in Publication Data
A catalogue record for this book is available from the British Library

ISBN (10): 1-4438-4947-2, ISBN (13): 978-1-4438-4947-0

TABLE OF CONTENTS

LIST OF FIGURES

List of Tables

INTRODUCTION

LORRAINE LOCKE AND SIMON LOCKE

This book grew out of a Science and the Public Conference that has become an annual event in the UK over the past several years. The 6th Annual Conference in 2011 was held at Kingston University in South-West London and billed as "A Quarter-Century of PUS" to mark, if somewhat belatedly, the 25th anniversary of the publication of the Royal Society's 1985 report into the public understanding of science (PUS) (aka the Bodmer report). Against this background, the Conference was seen as an opportunity to reflect on the development of the field over its brief life-time and consider its prospects for the future. The present book continues this intent by aiming to contribute to the movement of scholarly work beyond matters of deficit, engagement and transfer principally through a strong focus on what might be called "informal science education" (see Lewenstein's chapter below).

Much of the debate about PUS concerns the relationship between knowledge, education and democracy voicing a longstanding concern in western societies over the extent to which effective citizenship requires appropriate education. Can people be proper citizens if they have not been taught how to be so? On the other hand, is educating people in how to be proper citizens merely a guise for ensuring they continue to support the status quo? What, and whose, notion of "proper" is being advocated? Similarly, by "education" do we mean only that provided by formally designated educational institutions? Or is it possible for people to become self-educated "informally", outside of such formal contexts?

The tension over whether people need "proper" direction or if they are capable of directing their learning for themselves is apparent in the debate over PUS that ensued after the publication of the Bodmer report in the UK. Bodmer's clear view was that the public needed direction. Despite evidence of strong public interest in science and a distinct absence of evidence of widespread public ignorance, Bodmer emphasised a view of the public as needing "improvement" in their understanding of science and in so doing established the basis of what soon came to be called the "deficit model" of PUS. For Bodmer, this was not just a deficit of knowledge, but also one of democratic capability; because of the extensive presence of science in more and more areas relevant to public policy, having knowledge of science and its procedures was presented as being increasingly vital to effective citizenship. If people do not understand the science (and technology) about which policy decisions are being made, then how can they be trusted to make the right decisions, both indirectly via the ballot-box and more directly via the marketplace (a point addressed in Toumey *et al.*'s chapter below). Thus, to be proper citizens they need educating in science, but it has to be science of the proper kind. The anxiety expressed by Bodmer was that, if left to their own devices, the public are not merely effectively disenfranchised by their ignorance, but even worse, vulnerable to various kinds of *improper* "science", in fact not really "science" at all, but "pseudo-science" spouted by various kinds of charlatan in a growing threat from a New Age of superstition and anti-enlightenment–an "anti-scientific" attack on the very foundation of modernity (Holton 1992).

The deficit model quickly came under strong attack, partly for its over-simplification of a complex social world made up of multiple scien*ces* with multiple understan*dings* by multiple pub*lics* (Birke 1990; Ziman 1991) and partly because it was seen as a guise for advocating continuing support for the formal institutional apparatus of science: by "understanding" was really meant "appreciation" and "acceptance" (Wynne 1992). Likewise, the notion of "anti-science" was seen as a means of deflecting and undermining public criticism of the deficiencies of this formal apparatus. From this view, the deficit resided not so much with the public as with scientists, who themselves failed to appreciate ordinary people's knowledge or accept the legitimacy of their concerns about science and technology, thereby appearing arrogant and dismissive. Thus, it is not so much the public who need to be educated in science, as scientists who need to be educated about the public and, likewise, the democratic deficit resides in the institutional apparatus of formal science. From this viewpoint, the public do not require direction by scientists, or at least not by scientist-advocates of the formal order; rather, the public are both capable of self-direction in educating themselves about science and equally capable of deciding when such education is necessary for them–and indeed, when it is not (Michael 1996).

The upshot of this debate in the UK was a shift in language from "deficit" to "dialogue" (House of Lords 2000). In practice, however, the prevailing voice in science-public relations has become that of "engagement", but the extent to which this marks a genuine shift beyond the monologue of the deficit model has been called into question (Lock

2011). On the face of it, people's participation in discussion of the activities and still less the institutional orders of formal science seems rather less sought after than their enjoyment of interactive sideshows. Accordingly, the issue of democratic deficit has again been a focus of critical attack (Wynne 2008). From this perspective, "engagement" is an avoidance tactic: a way of appearing to address concerns over public involvement in science, while actually not changing anything fundamental at all. Indeed, engagement is really "deficit" by another name, since the directing concern remains that of improving public understanding (read: acceptance) of science albeit that the strategies employed to do so are along the lines of "edutainment" rather more than the formal didacticism of the professorial voice. Not so much "citizen science" (Irwin 1995) then, as consumer science.

Alongside this sop to citizenship has come a further development at in the form of the notion of "knowledge transfer" (KT). This term, rooted in organisation and management theory, pre-dates the broad adoption of "engagement" (one of us first encountered it as an Economic and Social Research Council PUS Fellow in 1998), but has since been taken up in the UK especially in regard to university research, which is now to be assessed for its KT "impact". This seems to cover a range of possible relations with wider publics, although perhaps with a preference for economic actors since KT is linked to notions of the "knowledge economy", thus emphasising the role of knowledge in directing innovation and production. In this respect, "transfer" is at the other end of the economic chain to "engagement" and taken together they might be seen as partners in a dual strategy to direct economic action principally through science-driven technological innovation. It is, after all, one thing to look to science as a source of new products, but quite another to ensure there is a ready supply of eager consumers for those products. If and in so far as developed economies now do look to science as the basic engine-driver of economic growth, then its legitimacy needs to be upheld or the products themselves may be seen as part of the problem rather than the solution. Thus, encouraging citizens to participate in genuine debate about the formal institutions of science is potentially perilous, since they may call into question the steering of scientific research toward currently preferred agendas of technoscientifically-constituted economic activity.

Viewed in this light, KT arguably shares with "engagement" the same basic framework of deficiency. Although there is much emphasis on recognising the grounding of knowledge in local "networks", nonetheless, the metaphor of "transferring" conveys the sense of shifting a given thing from one context to another in an essentially unchanged form.

Accordingly, knowledge is apparently invested with a thing-like quality, as something that can be simply reproduced from one situation to another, as in, for example, notions of transferring "best practice", carrying the implication that practices are divorceable from the immediate social contexts in which they are constituted and that they constitute. Such a conception of knowledge as a clearly bounded, thing-like given that can be lifted from one context and placed into another without change, informs the "knowledge-quiz" surveys that arose from the deficit model and came to epitomise its major flaws (Evans and Durant 1989; Irwin and Wynne 1996). KT indicates that, at least amongst those responsible for governing publicly-funded scientific research, there is little real interest in encouraging genuine public dialogue and rather more in imposing a restrictive regime of knowledge production in the hope of revivifying a flagging economy, the barbaric (in Matthew Arnold's sense) assault of the current UK coalition government on the social sciences and humanities being perhaps a more obvious indicator.

The papers in this book, however, point in varying ways and to different degrees to a significantly different conception of knowledge and its public(s). As stated, broadly these studies fall into what, following Lewenstein, we will call "informal science education", incorporated into which, as he outlines, are such outlets as museums, theatre, the mass media and fiction involving science. Accordingly, we have divided the papers into three parts reflecting their concern with: knowledges in publics; public venues of knowledge (museums and theatre); and public mediations of knowledge (mass media and popular culture).

Part One: Knowledges in Publics

The two papers in Part One provide summary overviews of the general state of discussion at the present time regarding how best to conceptualise the relationship between the public and science. Although both document discussions held in the United States, the issues they address have some general applicability, as indicated by the reference they make to the Bodmer report, but also as becomes apparent from ensuing chapters in which aspects of the global reach of the science-public relation are documented and explored.

As Chris Toumey *et al.* make apparent, these matters are of longstanding interest in the United States, which has had a tradition of survey research into "science literacy" reaching back over several decades. However, while acknowledging the longitudinal value of this work, their remit arose from recognition by the National Science Board of the

National Science Foundation that there was a need to revise the existing approach, which narrowly construed "science literacy" in terms not unlike those of the UK and European deficit model. Even by its own criteria of democratic efficacy, Toumey *et al.* consider this approach flawed, partly because it is not clear that high levels of science literacy are either attainable or necessary for effective democratic action and partly because the politically motivated conception of why people may want scientific knowledge is too circumscribed. Instead, Toumey *et al.* propose a framework for assessing public knowledge of science that adds two further interests to this "civic" motivation: "practical individual decision-making"; and "cultural curiosity" about science. In addition, they suggest that what is meant by "scientific knowledge" can cover three broad types of content: "scientific facts"; "processes and standards"; and "scientific institutions". Combining these produces a three by three framework of "purposes and content" from which to begin to develop a more sophisticated assessment of public knowledge of science conceived in the general manner of "science in the service of citizens and consumers" (consumers, that is, of knowledge).

Despite misgivings in some quarters as Toumey *et al.* record, the framework has been welcomed by the British Science Association and promises to make a significant advance over existing survey-based approaches, so we are very pleased to be able to present it here for wider European audiences. As the team also acknowledge, however, it remains restricted to assessing scientific "knowledge" and not "interpretations" of this knowledge, a distinction that is not necessarily clear-cut as some of our later chapters show. Nonetheless, one of the strengths of the framework is that it attempts to recognise the multiplicity of the public's relationship to science and the diversity of sources of knowledge that self-directed learners may employ. Moreover, in incorporating some measure of "institutional knowledge", additional recognition is given to the social context within which scientific knowledge is generated and applied. In these respects, there is strong continuity with Bruce Lewenstein's following chapter on "informal science education" (not altogether surprising, given his involvement with both reviews).

Lewenstein summarises a report on *Learning science in informal environments* commissioned by the National Research Council of the United States National Academy of Sciences, but also uses this as an opportunity to reflect a little on what he suggests can be considered a "[natural] scientific" approach to PUS. The reason for this is because the report was subject to peer review, which brought out issues over different standards of evidence employed by natural scientists vis-à-vis social scientists and humanities scholars; as Lewenstein neatly sums it: "'data' is not the plural of 'anecdote'." Consequently, much of the case study research found in science and technology studies–the sort of research, in fact, that makes up the bulk of this book–might be deemed of insufficient validity by the NRC, however persuasive social scholars of science may find it. Given that, from Lewenstein's account, this includes much of the work on informal science learning (i.e., the sort of learning that occurs in contexts outside of formal education and related institutions), it can perhaps more readily be appreciated that natural scientists may be unwilling to count such learning as really "science". On the other hand, we might also turn this around and view the resort to such methodological matters as a standard technique of boundary-work (Gieryn 1999), long recognised in science studies as an exclusionary tactic (Collins and Pinch 1979). Thus, in pointing out this issue, Lewenstein makes us aware that such defensive manoeuvres are available for deployment even in what he calls a "backstage" arena. Nonetheless, he ends on a hopeful note, concluding that it is important to sustain the conversation across the disciplinary boundaries and that the field of science and the public has great significance as a meeting place.

Lewenstein highlights a number of features of the report itself, which identifies a range of "venues" of informal learning (such as museums) and "themes" that inform such venues (such as media), as well as offering a way of thinking about "learning" as a weaving together of six "strands" that are not restricted to formal educational contexts. Amongst these, he points especially to reflection on scientific institutions as a significant inclusion for the involvement of science and technology studies, using as an illustrative example the case of museums as both major cultural institutions and venues where "personal interaction" between visitors and staff can occur. These features significantly figure in the two chapters that follow.

Part Two: Public Venues of Knowledge

In Parts Two and Three of the book, we move from general theoretical and conceptual discussions to specific examples of empirical research. Part Two includes studies of two major kinds of venue in which informal science is displayed and performed: museums and theatre.

In chapter three, Claudia Geyer, Katrin Neubauer and Doris Lewalter report on their evaluation survey of visitors to a number of museums across Europe involved in the EU Commission funded NANOTOTOUCH project. The project aims to make the scientific laboratory directly accessible to the public by setting up nano-research facilities within museums and encouraging visitors to interact with the researchers. Geyer *et al.* focus on assessing the effectiveness of the project in terms of how well it succeeded in generating and sustaining visitors' "situational interest" in nanotechnology and their self-perceived increase in knowledge about nanotechnology and research. Amongst the key findings is that the perceived relationship to the scientist is closely related to both these factors: the better the perceived relationship, the better the reported experience and learning. This supports Lewenstein's point that museums are particularly well-positioned to take advantage of the personal interaction with staff scientists their venues afford.

The other feature of museums Lewenstein picks out is their significance, not merely as venues for "knowledge transfer" narrowly defined, but also as national and cultural heritage sites with the potential to locate science and technology in wider social contexts. This is developed in chapter four, Ines Hülsmann and Ana-Maria Raus's study of displays of the pacemaker. Drawing on actor-network theory and so providing a qualitative counterpart to Geyer *et al.*'s quantitative analysis, Hülsmann and Raus describe how pacemaker technology is represented in three different Dutch venues, a museum, a science centre, and a manufacturing company. Their descriptions clearly show how ostensibly the same "object" is quite differently configured in these different sites demonstrating that there is no simple "transfer" of knowledge, but a distinct "translation", which in each case offers quite different relations to visiting publics. This is made all the more apparent by the fact that the science centre works directly with the manufacturing company and yet displays the technology in a significantly different way, to the point that one company interviewee expresses concern that the centre provides not "edutainment" but merely "entertainment" (evidently then, it is not just social scientists who worry about the "engagement" agenda, even if for rather different reasons). In a notable point of confluence with Geyer *et al.*, Hülsmann and Raus also highlight the importance of the interaction between venue staff and visitors, and they echo Lewenstein in proposing that all the venues could do more to increase awareness of the social in science and technology.

One venue where such awareness is being articulated, at least to some degree, is theatre and our next two chapters provide examples of two rather different approaches to accomplishing this, which despite contrasting starting points, show significant convergence towards similar outcomes. In chapter five, Jeff Teare provides a summary overview of the work to date of Theatrescience, a drama company that has become increasingly involved in developing and exploring ways of using drama to address moral and ethical issues involving science. From the outset, Theatrescience's work involved more than simply putting on plays; rather, with the support of bodies such as The Wellcome Trust and the Eden Project in the UK, they have explored ways in which to bring together drama writers and practitioners with scientists and publics in schools, at science centres and in wider communities. As part of this, they have sought to break down longstanding barriers between the arts and science and, in this respect, provide a practical example of the kind of interdisciplinary conversation afforded by the field of PUS. They have also worked internationally, notably in India and Africa, on developing dramatisations of health and related issues that involve direct input from affected communities and it is here especially that their work shows convergence with that of Aiding Dramatic Change in Development (ADCID).

In chapter six, Stephen Sillett and Jennifer Jiminez provide an account of ADCID's work involving Socio-Drama Topography (SDT). This is described as a "reflective and dialogical arts-based inquiry process" that seeks to use dramatic techniques developed with the active involvement of communities to air local issues with a view to encouraging reflection and greater empowerment for participants. The two applications of SDT outlined in the chapter are concerned with health and environmental issues in deprived areas of rural South Africa in which western scientific understandings are crucially positioned. In the first case concerning HIV/AIDS, from which the SDT approach was first developed, the project was concerned with improving young people's adoption of safe-sex practices and involved learning from both the youth and their parents about traditional views of sexual relationships and their importance to identity. Drama was then used to facilitate community discussion and reflection with a view to developing knowledge and changing practices towards "healthy choices". In the second case, focused on issues of water access and hygiene, SDT was employed from the outset to assist a community-based organisation seeking to change from a more technically-driven mode of service provision towards greater community partnership. As Sillett and Jiminez put it, this "involved a shift in engagement from transferring knowledge and servicing needs to developing the 'power to empower'." In short, a shift from top-down to bottom-up and from formal, institutionally-directed knowledge to informal, locally-grounded knowledge.

Thus, despite their rather different starting points, both Theatrescience and SDT have developed forms of drama-based intervention that attempt to encourage participatory involvement of local communities and to incorporate their knowledge(s) into the learning activity. There is about this a clear resonance with established critiques of the deficit model that also emphasise the importance of local knowledge(s), a matter that arises again, albeit in rather different fashion, in Part Three.

Part Three: Public Mediations of Knowledge

In Part Three, we shift locus from what might once have been considered "high culture" venues to the mass media and popular culture, beginning with two chapters looking at science and scientists in newspapers and magazines.

In chapter seven, Miquel Carandell provides a detailed analysis of the fascinating controversy in the Spanish press over a fossil bone fragment referred to as "Orce Man", with respect to its classification as indeed a hominid or merely a donkey. The matter was controversial, because its acceptance as hominid presented a challenge to the prevailing view of when early humans arrived in Europe. Carandell uses the case to assess the applicability of a number of models of science popularisation and to highlight that newspapers may themselves become a site for establishing scientific knowledge (referred to as "medialisation"). In this way, the traditional sharp division between formal contexts of scientific knowledge production and the informal context of popularisation is called into question, as it is argued that for at least some members of the scientific community, newspaper reports provided the primary source of information regarding the nature and status of the fossil. Further, not all the scientists involved were distant from the controversy, but some closely involved. The case then provides reinforcement for the view that learning about science is something that can occur in a range of contexts, whether considered "formal" or "informal" and that this is not just true for non-scientists. Members of the scientific community may also rely on "informal" sources, troubling the distinction between scientists and the public and raising a question about how we should view these categories themselves.

One proposal in response to this question comes from Simon Locke in chapter eight, in the form of a preliminary application of Membership Categorisation Analysis (MCA) to some treatments of academics and doctors in newspapers and magazines. MCA is a perspective associated with Harvey Sacks (better known for having founded conversation analysis), which over recent years has become increasingly widely used to study the ways in which categories are employed by people in everyday life to make sense of their own and others' activities. Actions can be matched to categories and categories to actions providing us with a ready interpretative apparatus to understand what is going on in a given situation. Arguing that category usage involves rhetorical reasoning conforming to the incomplete syllogism of the enthymeme, Locke provides illustration of how non-scientists writing in the mass media can use the attributes or "predicates" associated with the categories "academic" and "doctor" to construct critical accounts of science and scientists. In particular, he notes that academics may be judged "bad" in relation to a set of norms of a broadly Mertonian form that are often found in popularised representations of scientists. Thus, scientists are held publicly accountable to a moral order they have themselves worked to constitute in public. Similarly, from consideration of contemporary press reports of Jack the Ripper, Locke suggests that members of the public may draw on scientific knowledge to construct versions of "madness" to which scientists are then held accountable, even as actions attributed to the category of "scientist" are judged by the existing moral order, in particular acting instrumentally. Finally, Locke also gives some brief consideration to the additional category "local" to point out that the use of this common sense construct by social scientists as a resource with which to criticise the deficit model demonstrates our own reliance on ordinary social-rhetorical reasoning, further troubling any easy distinctions between scientists (natural or social) and the public, and between formal and informal knowledge.

Such questions arise also in relation to fiction involving science in which, in a variety of ways, the boundary between "fact" and "fiction" becomes distinctly (and oxymoronically) fuzzy and our last two chapters carry this matter forward in two novel (pun intended!) directions. In chapter nine, Alice Bell looks at the classic British children's character, Professor Branestawm, as an example of how images of the scientist may be informed by (modern western) cultural representations of the child. Bell points out that there are a variety of contrasting qualities associated with children, such as being innocent or corrupt, that may also be associated with scientists and enable positive or negative images of the latter to be constructed through the former. In the case of Branestawm, she suggests that a range of positive qualities of the child are combined with "anti-childlike", "mature" features that enable different versions of the character to be presented, which she tracks over the course of his 50 year history (from 1931-1983). She further suggests, however, that such negotiations of the "child/scientist boundary" are not just to be found in the fictional character, but also amongst "real world" scientists, "to take advantage of both positive qualities of childishness as well as the advantages of appearing distinctly 'mature'." Thus, through studying such fictional characters, we may learn not only much about the common sense attributions associated with scientists, but also perhaps, those with which scientists may seek to associate themselves. This then is one way in which the "informal" context of popular fiction may blur over into more "formal" scientific self-representations.

Bell's discussion points to the "ambivalence" with which scientists are represented in children's literature, a matter that comes through also in our final chapter. Here, Kate Roach raises the intriguing question as to why the "scientific detective" has been largely excluded from discussions of fictional representations of the scientist, which tend to focus over-narrowly on "mad" scientist figures, themselves often viewed as one-dimensional critiques of the "evils" of science. Through detailed consideration of some of the most influential early detectives in 19th century popular fiction, she argues they were represented as "men of science", but also incorporated "occult" qualities that, during the 20th century, became marginalised from formal science. Yet, in popular culture, such figures retain an important presence, most prominently in Dracula's nemesis, Professor Van Helsing, both scientist and supernaturalist. Together with the more widely studied (in the PUS literature) mad scientist proto-type, Doctor Frankenstein, Van Helsing makes up something of a double-act that looms large over contemporary popular images of the scientist and from which a host of fictional progeny has spawned. Roach suggests that this supports the view that the public meaning of science remains mixed and uncertain, combining both enchanted and disenchanted qualities that are played out and explored through such images. Popular culture, then, is a site where the meaning of science is debated in part through such narratological resources and the implications of this for both "formal" and "informal" science education are much in need of further study.

Taken overall, this is the message of the book: that the need remains to move the agenda of PUS beyond matters of deficit, engagement and transfer, all of which despite the differences in their packaging prioritise the position of formal institutional science as the primary source of knowledge production in our society. Our commitment, in contrast, is to a view of knowledge as something that pervades all parts of the social world, as a stock of sedimented typifications, discursive practices and artful accomplishments that underwrite and make possible all forms of social life including those we call the sciences. In the first instance, science grows from the everyday lifeworld (for which the term "public" is a certain kind of synonym) and though, as it feeds back into it the lifeworld may become transfigured, science is itself transmogrified in the process. Such transformations, however strange and unusual they may seem and however "informal" their constitution, require empirical study and understanding more than anxious condemnation or self-interested re-direction. Publics *do* things with science, they generate their own knowledges in accord with their own concerns and criteria, regardless of the needs and desires of any formal, institutional apparatus: get used to it.

References

Birke, Linda. 1990. Selling science to the public. *New Scientist*, August 18: 40-44.

Collins, Harry M., and Pinch, Trevor. 1979. The construction of the paranormal: nothing unscientific is happening. In *On the margins of science: the social construction of rejected knowledge*, ed. Roy Wallis, 237-270. Keele: Keele University Press.

Evans, Geoffrey, and John Durant. 1989. Understanding of science in Britain and the USA. In *British social attitudes: special international report*, eds. Roger Jowell, Sharon Witherspoon and Lindsay Brook, 105-119. Aldershot: Gower.

Gieryn, Thomas F. 1999. *Cultural boundaries of science: credibility on the line*. Chicago: University of Chicago Press.

Holton, Gerald. 1992. How to think about the "anti-science" phenomenon. *Public Understanding of Science* 1(1): 103-128.

House of Lords. 2000. *Science and society, report of the select committee on science and technology*. London: HMSO.

Irwin, Alan. 1995. *Citizen science: a study of people, expertise and sustainable development*. London: Routledge.

Irwin, Alan, and Brian Wynne, eds. 1996. *Misunderstanding science? The public reconstruction of science and technology.* Cambridge: Cambridge University Press.

Lock, Simon J. 2008. *Lost in translations: discourses, boundaries and legitimacy in the public understanding of science in the UK*. PhD diss., University College London.

Michael, Mike. 1996. Ignoring science: discourses of ignorance in the public understanding of science. In *Misunderstanding science? The public reconstruction of science and technology*, eds. Alan Irwin and Brian Wynne, 105-125. Cambridge: Cambridge University Press.

Royal Society 1985. *The public understanding of science*. London: The Royal Society.

Wynne, B. 1992. Public understanding of science research: new horizons or hall of mirrors? *Public Understanding of Science* 1(1): 37-43.

—. 2008. Elephants in the rooms where publics encounter “science”?: A response to Darrin Durant, “Accounting for expertise: Wynne and the autonomy of the lay public”. *Public Understanding of Science* 17(1): 21-33.

Ziman, J. 1991. Public understanding of science. *Science, Technology, & Human Values* 16: 99-105.

PART ONE

KNOWLEDGES IN PUBLICS

CHAPTER ONE

RETHINKING PUBLIC KNOWLEDGE OF SCIENCE: THE PROCESS OF CRAFTING THE CONCEPT OF "SCIENCE IN THE SERVICE OF CITIZENS AND CONSUMERS"

CHRIS TOUMEY, JOHN BESLEY, MEG BLANCHARD, MARK B. BROWN, MICHAEL COBB, ELAINE HOWARD ECKLUND, MARGARET GLASS, THOMAS GUTERBOCK, ANTHONY E. KELLY AND BRUCE LEWENSTEIN

Introduction

The *Science and engineering indicators* (hereafter *Indicators*) is a massive compilation of data that is assembled by the staff of the United States National Science Foundation (NSF) and published biennially by the National Science Board (NSB). Chapter Seven of the *Indicators* is titled "Public attitudes and understanding". When the 2010 edition was being prepared, some members of NSB criticised the item on public knowledge of evolution: "Human beings as we know them today developed from earlier species of animals". This item, they asserted, failed to distinguish between *knowledge of* evolution and *belief in* evolution (Bhattacharjee 2010). A person could know that scientists say that humans have evolved, but still disagree with the scientists; in that case, which is sometimes detected in survey research, the interviewee knows how evolution is presented by scientists or the authors of textbooks, but at the same time he or she does not believe in evolution. In the view of some members of NSB, the wording of the item on evolution captured *belief in* evolution when it should have captured *knowledge of* evolution instead.

After a series of communications were exchanged between NSB members and NSF staff, it was agreed that the conceptual framework for public knowledge of science, as reported in Chapter Seven of the *Indicators*, ought to be reexamined. A workshop at NSF was planned to reevaluate the conceptual framework in October 2010, with a follow-up workshop to devise methods to implement the recommendations of the first in November 2010. The authors of this paper participated in the first workshop, and some members also contributed to the second workshop. Here we describe the process of: (1) examining the former conceptual framework; (2) suggesting a different framework; (3) clarifying the implications of the second framework; (4) observing how the second framework was incorporated into documents of the National Science Board; and (5) observing how the second framework was reported in science media.

The topic of public knowledge of science deserves a rich interdisciplinary approach. The participants for the workshop of October 2010 had expertise in science communication, science policy, science education, informal science education, survey design, learning-and-cognition, science-and-culture, and other related areas. That way the group could look critically at public knowledge of science from multiple relevant perspectives.[1]

From Civic Scientific Literacy to "Science in the Service of Citizens and Consumers"

Why should the National Science Foundation measure public knowledge of science, and why should the National Science Board publish this information? These were the initial questions that the workshop considered. The workshop participants noted that NSF and other governmental science agencies have a legitimate interest in knowing how the public examines scientific evidence, how the public reasons about evidence and how it uses evidence to make judgments either as individuals or as communities. In the words of NSF's (2003) strategic plan, one of its objectives is to

> promote public understanding and appreciation of science, technology, engineering, and mathematics, and build bridges between formal and informal science education.

For purposes of conceptual clarity, the workshop participants used the term "public knowledge of science" for three reasons. First, there was concern that the expression "public *understanding* of science" has acquired a highly charged negative connotation in both the research and the policy communities as a result of criticism of projects conducted earlier under that title. This problem arose after the Royal Society presented its 1985 report, *Public understanding of science*, also known as the Bodmer report (Royal Society 1985). This document has been widely diagnosed as a plan in which scientists talk, members of the public listen and then the public uncritically supports government funding of scientific research. An opposition to that plan quickly crystallised, as represented in Brian Wynne's (1992) paper, "Public understanding of scientific research". There, the author asserted that the Bodmer report was motivated by scientists' selfish fear of losing public support for science. In the words of Wynne (1992, 42), this reflected "the social neurosis of science over its authority and public legitimation", in which the work of scientists is not vetted by the public. Wynne (1992, 37) writes that "problems in public understanding of science reflect problems in the representation, organisation and control–the broad political culture–of science". This and other critiques have painted the Royal Society report as misguided and unrealistic. We note that Sir Walter Bodmer recently defended the report, saying that critiques have oversimplified its conclusions (Bodmer 2010).

Second, the conceptual framework to be reevaluated, public knowledge of science, is often identified with the term "civic scientific literacy". If hypothetically the workshop was to recommend a different conceptual framework, then the themes of the new framework would lead to a new terminology. Third, "understanding" can include both the scientific knowledge that the public possesses and the attitudes, values, concerns, perceptions and other factors that shape public interpretations of that knowledge.

The workshop participants were charged to reevaluate the conceptual framework for public knowledge of science, but not the influences that shape interpretations of knowledge. Those other influences are interesting and important, but the problem at hand was public knowledge of science. Furthermore, a reevaluation should think about the future: how can a conceptual framework improve the process of measuring and reporting information for the 2014 *Indicators* and beyond?

The first order of business of the workshop was to examine the history of measuring and reporting public knowledge of science. Dr Robert Bell of the Science Resources Statistics Division at NSF (subsequently renamed as the National Center for Science and Engineering Statistics) presented this history from an administrative perspective, after which the workshop participants discussed the contributions and conceptual framework of Dr Jon D. Miller, who established a framework known as civic scientific literacy (CSL) in 1983, with various revisions since then (Losh 2006).

Miller's framework was anchored in John Dewey's theory of liberal democracy, particularly Dewey's 1934 essay on "The supreme intellectual obligation" (Dewey 1981 [1934]; Miller 1983, 1987a, 2004). Here, Dewey argued that if citizens know how to think scientifically, then democracy will benefit from good knowledge combined with good decision-making processes. According to Miller's (1983, 29) account,

> In a democratic society, the level of scientific literacy in the population has important implications for science policy decisions…any measures we can take to raise this level…will improve the quality of both our science and technology and our political life.

None of the workshop participants opposed civic scientific literacy *per se*. Nevertheless, they identified two reasons to develop an updated conceptual framework. One is that the former vision has not been attained. It is possible that higher levels of scientific thinking might or might not affect democracy for the better, but there is little reason to be optimistic that the American public will achieve the levels of scientific literacy that Dewey and Miller hoped for. The civic virtue that Dewey envisioned included individuals voting and making personal decisions. Some readers might further infer that Dewey also called for the kinds of large-scale political grassroots organising that are required to support or resist a particular science policy. Even so, telephone surveys have not captured that latter possible dimension of civic scientific literacy. It can be recognised that large-scale political activism is now a common feature of public scientific controversies in creation-evolution disputes, AIDS/HIV policy, environmental issues, and other topics. That level of activism on scientific topics proceeds with or without desirable levels of scientific literacy. A conceptual framework for public knowledge of science should reflect the reality that scientific knowledge is acquired and deployed, not only in voting in elections and referendums, but also in additional styles of civic engagement.

The second reason for reevaluating the conceptual framework of civic scientific literacy is that this vision frames the person in the public as a micro-scientist. That is, it identifies some of the knowledge that working scientists possess and then measures how much of that knowledge non-scientists possess. Consistently the answer is that most of the public possesses miniscule quantities of scientific knowledge, leading to stories with titles like "America's scientific illiterates" (Russell 1986), "The dismal state of scientific literacy" (Culliton 1989), and "The scientifically illiterate" (Miller 1987b). The workshop did not challenge the validity of these reports.

What should be the standard of acceptable civic scientific literacy? Sometimes it is said to be the ability to read the "Science" section in the Tuesday edition of the *New York Times*. Why? If a citizen accepts that scientific information passively or uncritically, is this an acceptable form of civic scientific literacy?

The workshop participants agreed that decades of data collection from surveys of civic scientific literacy have enabled high-quality longitudinal research. Long-term trends can be identified and analysed. Likewise, comparative research is made possible. Public knowledge of science in the United States can be weighed against the same in other nations and perhaps insights can be derived from that kind of comparison. This kind of analysis is already made possible for K12 science education, e.g., in the *Science framework of the 2009 national assessment of education progress* (NAGB 2009). It would be regrettable if the longitudinal and comparative value of that information was diminished.

Following that conclusion and with the benefit of the participants' expertise in science communication, science policy, science education, informal science education, survey design, and other related topics, the workshop explored ways to improve the conceptual framework by incorporating recent thought about relations between the science and the public. One insight that was especially salient is that persons in the public have different reasons for acquiring scientific knowledge and using it (e.g., Bell *et al.* 2009; Shen 1975; Toumey 2006; Wickson *et al.* 2010).

Sometimes a person is in the role of an information consumer and so wants the kind of practical knowledge that enables one to comprehend the ingredients in a food label, or to know how to take antibiotics without developing antibiotic-resistant bacteria. Other times a person is in a civic role and needs scientific knowledge in order to have an active and constructive role in a science policy decision-making process. If a nuclear reactor is planned near one's home, what knowledge will a person need to weigh the benefits and the risks, and then to participate in supporting or opposing the construction of the reactor? In a third situation, a person might feel that science is interesting and learning about science is enjoyable. Unlike the reasons of the consumer or the citizen, this motive has merely the pleasure of learning about science. We can call this public knowledge of science for its own sake and we can note that by acquiring it, people are connected to a shared view of how the natural world works.

In addition to considering the reasons *why* people acquire scientific knowledge, it is worth realising that there are *different kinds* of knowledge and that some kinds will serve one purpose while others serve another. The consensus of the workshop was that there are three principal categories of scientific knowledge that can serve persons in the roles of information consumers, citizens, and the curious:

1. *Factual scientific knowledge* gives one a vocabulary of scientific information and scientific conclusions about the empirical world. For example: What is an atom? What is a species? What is a vitamin? What are genetically modified organisms? What are stem cells? In addition to knowledge that might be conveyed as definitions, it also includes natural and technical processes: What is adaptation, and how does it work? How does a solar cell work? How does a nuclear power plant work?

2. *Knowledge of scientific processes and standards* enables one to comprehend intellectual practices such as experimental design, naturalistic explanation, sampling and probability, and so on.

3. *Institutional scientific knowledge* enables one to know how scientific institutions operate. This includes peer review; the adjudication of scientific claims; the funding of scientific research; how science identifies and prioritises emerging issues; how scientific advice is used; processes of making science policy; and so on.

From those considerations comes the core of a conceptual framework for measuring and reporting public knowledge of science in the *Indicators*:

> In order to place science in the service of citizens and information consumers, the concept of public knowledge of science refers to: (a) factual scientific knowledge; (b) knowledge of scientific processes and standards; and (c) knowledge of how scientific institutions operate. It equips persons in the public for: (1) active civic engagement in scientific issues, including organised efforts to support or oppose specific science policies; and for (2) using scientific knowledge for practical decision-making by individuals; and for (3) a better scientific understanding of the world.

In addition, the process of measuring and reporting public knowledge of science continues the long-term responsibility of collecting data which enables high-quality longitudinal and comparative analysis.

This conceptual framework can be envisioned as a three-by-three matrix. The horizontal dimension represents three purposes for acquiring knowledge, and the vertical dimension depicts three kinds of knowledge content. One can then categorise items to be measured according to which purpose they serve and what kind of content they represent (Table 1.1).

		Purposes of public knowledge of science		
		Civic engagement with science	**Practical/ individual decision-making**	**Cultural curiosity about the scientific worldview**
Content	**Factual knowledge**		How should antibiotic medicines be used?	What is an electron?
	Processes and standards	How is probability relevant to a particular issue?		Principle of naturalistic explanation
	Institutional knowledge	Why does nano-technology receive government funding?	Which experts and institutions can I trust?	

Table 1.1. A 3x3 matrix of Purposes and Content showing how certain kinds of knowledge fit into cells

For example, the principle of naturalistic explanation would belong in the row for scientific processes and standards and the column for scientific understanding of the world. It would also go in the column for the civic purpose of public knowledge of science in the case of a policy controversy about evolution and creationism. But it is not necessarily urgent for it to be in the column for the practical purpose of serving consumers. One can imagine how a person who wants to understand the label of ingredients on a food package does not particularly need to invoke the standard of naturalistic explanation. It is noted that some items to be measured can go in more than one column and more than one row.

This matrix can be further understood by focusing on one particular theme. In this case, we place nine kinds of knowledge about medications into the matrix (Table 2.2).

		Purposes of public knowledge of science		
		Civic engagement with science	**Practical/ individual decision-making**	**Cultural curiosity about the scientific worldview**
Content	**Factual knowledge**	Who funds research on Drug X?	What are the risks of Drug X for me and my family?	How do drugs work?
	Processes and standards	What standards are used to evaluate Drug X?	Have financial interests affected safety testing of Drug X?	How do scientists develop drugs?
	Institutional knowledge	How can non-experts affect research and regulation of Drug X?	How can I evaluate conflicting reports about Drug X?	What is the social history of research on Drug X?

Table 1.2. A matrix of Purposes and Content focusing on questions about medications

The starting point of this conceptual framework is to ask what knowledge a person in the public needs, whether for civic engagement with science and science policy, or for making individual decisions about one's life or health, or for feeding one's curiosity about science. This starting point is different from that which informed the previous conceptual framework, when the principal effect was to measure civic scientific literacy as a proportion of scientific knowledge in general (and how little the public knows). The revised framework entails a series of consequences for how we think about relations between the public and scientific knowledge. The workshop participants noted that this was not the original intent of the framework of civic scientific literacy, which had a complex definition of scientific literacy that included both science process and science policy (Miller 1983). In practice, however, that framework was largely understood as focusing on the science content dimension. The revised framework, "Science in the service of citizens and consumers" (SSCC), entails a series of consequences for how we think about relations between the public and scientific knowledge.

A Conceptual Framework Based on Citizens' Needs

The public is not a homogeneous entity. There are various levels of formal education and multiple levels of encountering science through informal science education. Topics of interest will differ. Some people will be interested in nuclear power; others will concentrate on one disease or another; still others will be curious about the ethics of embryonic stem cell research; or what they need to know for a career in environmental management; and so on. Furthermore, some people will care about a given issue more than others. The first responsibility of those who disseminate public knowledge of science is to serve the segments of the public that want this knowledge. This takes precedence over an aspiration to deliver public knowledge of science to everyone equally, including those persons who do not particularly care about scientific knowledge.

Thus public knowledge of science is largely topical according to this framework. This can be contrasted with universal or timeless scientific principles. Topical knowledge does not arise from the same needs as the content in a science course or a science textbook. On the contrary, it arises when a citizen or a consumer is curious, concerned, alarmed, or excited about a particular topic. A resident of the Louisiana coast may want to know how the residue of the oil spill in 2010 can be made to disperse. The molecular structure of hydrocarbons is relevant at one level, but the resident probably does not want a tutorial on that. Instead, he or she wants to know which products will work, how quickly they will work and whether they will harm the coast.

Related to the topical character of public knowledge of science is the point that non-scientists can often acquire, comprehend and employ the relevant scientific knowledge when they have to. Self-motivated learning by adults has an impact almost as strong as formal undergraduate science courses (Miller 2004, 289-290). It is not expected that, during a controversy or a crisis, persons in the public will aspire to acquire knowledge equivalent to a degree in a scientific discipline. But these citizens do not need to become scientists with formal degrees in order to know what they need to know to have active and constructive roles in public debates that include a scientific dimension. This reinforces the insight that the starting point for public knowledge of science is the need of the citizen or the information consumer, rather than a microcosm of what a scientist knows. Consistent with this perspective, the workshop participants also recommended that periodically a topic of special concern in the United States, e.g., genetically modified crops, be featured in the *Indicators*, with a series of questions to gauge public knowledge of the subject.

Next, it is no secret that persons in the public, like persons in scientific communities, seek scientific knowledge from multiple sources. It is known from the 2012 *Indicators* that television and the internet are the two principal sources of scientific information for the American public, in equal proportions. Access to knowledge is not limited to a small number of authorities. When persons in the public acquire scientific knowledge from institutions and persons that are considered authoritative by the standards of scientific communities, those institutions and individuals are communicating in a very competitive marketplace where other sources claim to be equally authoritative.

The workshop participants understood that the new conceptual framework overlapped with civic scientific literacy in the data collection it recommended, but it would also be more encompassing than that earlier framework. By updating the framework to account for research and critiques generated in the last twenty-five years, the participants sought to retain the value of data collected under the framework developed by Jon D. Miller, while providing a more robust structure with new perspectives on public interactions with science. The new framework makes explicit some assumptions that were earlier implicit and it changes some of the emphases. By re-reading Miller's work on civic scientific literacy over the past thirty years, one could find parts of the new conceptual framework prefigured there. The fundamental goal of collecting data on public knowledge of science, namely to serve government policy making, remains the same.

In addition to specifying these implications, the workshop of October 2010 made a series of recommendations about collecting data on public knowledge of science for the *Indicators*. These can be found in the workshop report, "Science in the service of citizens and consumers" (Toumey *et al.* 2010).

Measurement and Operationalisation of the SSCC Framework

Issues of measurement and item validity have not gone unattended in this fertile field of social science inquiry. And yet the renewed controversy over validity of the items used in the NSF surveys called for a careful examination of the quality and utility of the full set of public science knowledge items drawing on recent advances in survey methodology. The second workshop, on implementing the recommendations of the first (Guterbock *et al.* 2011), examined the measurement adequacy of the current NSF survey items themselves taking into account the newly defined SSCC conceptual framework. The workshop, convened by Thomas Guterbock on 12 November 2010, brought together a group of survey methodologists and substantive experts for the purpose of developing a set of specifications to identify the measurement qualities that would be desirable in the public science knowledge questions and to outline a protocol for creating additional questions and testing them. The workshop participants represented a wide range of expertise from the disciplines of sociology, communication, psychology, political science, and health policy, plus survey researchers and methodologists.

Considerable scientific attention has already been paid to the assessment of the measurement properties of the existing science knowledge items used for the *Indicators*. This existing instrumentation assessment work is of high quality and was of considerable value to the evaluation task. Nevertheless, the workshop found that further study of some of the survey items is warranted and some new items will need to be developed if adequate measurement of the new framework is to be achieved.

Since the SSCC framework is broader in scope than its predecessor, the current NSF survey items do not measure all its aspects. Both factual and process items are well covered by the items in current use by NSF. However, it was noted that no relevant questions exist in the category of institutional scientific knowledge. Examples of this type of knowledge might be items asking about the federal government's role in funding basic research, the role of universities, differences in credibility of independently funded research versus that funded by for-profits, and so on. Another topic of institutional scientific knowledge was the question of human subjects in research. While funding agencies typically have clear parameters for protecting human subjects, this may be unknown to non-experts. It would be valuable to know whether this form of institutional knowledge can be detected and measured in survey research.

A key aspect of the SSCC framework is the recognition that citizens use scientific knowledge for three different sets of purposes as was indicated in the three columns of the matrix shown above. As the workshop reviewed the existing survey items it became evident that they do not attempt to measure these purposes directly. The second workshop found that it would be relatively easy to develop new items that would directly measure whether a person deploys scientific knowledge in everyday life. For example, items could probe whether the respondent regularly consults scientific or technical sources or published data in making important consumer purchases. Other items could measure whether the respondent holds opinions on policy issues where science is relevant and whether she or he relies on science knowledge in forming such opinions. A third set of items could focus simply on the extent to which a person enjoys hearing or reading about scientific studies, or learning about science.

Three distinct "science-purpose" scales could be constructed from these items. Armed with these scales, researchers could then put to empirical test the assumptions that underlie the SSCC framework, i.e., that science knowledge actually *does* empower citizens or improve their lives by making them more able consumers, more effective citizens, or better able to comprehend the world.

Although the instrumentation workshop had the task of reviewing the public science knowledge questions as a whole, the workshop participants devoted some attention to the two items that had drawn criticism from the NSB when the 2010 *Indicators* was being prepared: the true/false questions regarding evolution of humans from lower forms of life; and the origin of the universe in the Big Bang. While these questions are stated as simple factual propositions without any direct religious content, it is clear that some respondents respond to the items based on religious belief systems to which they are committed. In particular, conservative Christians who hold the Bible to be inerrant would be reluctant to endorse these items as being "true". The continued strength of conservative Christianity in the United States would then explain the lower scores on these items compared with other developed countries.

That presents an interesting problem. The evolution and Big Bang questions are different from the other knowledge items, but they also correlate fairly well with the other general-knowledge items. And so they function in part as measures of science knowledge, but they are also clearly picking up another dimension which may be personal commitment to religious belief. To derive the greatest analytical value from these items it would help to measure these two topics in terms of both knowledge and belief.

It was the strong consensus of the workshop participants that the notion of evolution is too fundamental and broad-reaching a concept in science to be left out of the set of public science knowledge indicators. Based on prior experiments conducted for NSF on variations of these items, the workshop suggested that they be modified into an "unfolding" or contingent form. Respondents would first be asked if the item were true "according to evolutionary theory" or "according to astronomers". They would then be asked directly if they personally share each belief. That approach would have the virtue of clearly separating a respondent's knowledge about what scientists believe from his or her personal beliefs.

The NSF public science knowledge items should be clearly focused on scientific knowledge. A change in measurement of the concept will not in itself resolve the debate over whether personal belief in human evolution is an essential part of scientific knowledge, or more generally whether science literacy necessarily assumes acceptance of scientific consensus as opposed to mere knowledge of that consensus. The suggested unfolding version of the questions will provide researchers with information that more clearly separates a respondent's knowledge of scientists' views from his or her personal agreement with those views, thus allowing for a more informed investigation of both aspects of public knowledge of science.

Looking to the research agenda ahead, the November 2010 workshop expressed concern that some of the current NSF survey items have been developed for use only in face-to-face interviews (although older items were developed for oral administration via telephone). The workshop recommended that a telephone-friendly version of current and proposed new items be developed, so as to allow more frequent and widespread testing of items by researchers who wish to expand, explore and validate the SSCC approach. The current survey vehicle for the NSF items is the General Social Survey (GSS): a biennial, NSF-funded, face-to-face survey known for the quality of its probability-based, large national sample. If research in this area is to develop at the needed pace, it will be necessary to collect a great deal of new data. A telephone version of the items would facilitate more rapid, more frequent and affordable data collection by multiple research teams to supplement and inform continued data collection via the GSS.

If the current measures are to be improved and the SSCC framework is to realise its full potential, there is thus much methodological work–

qualitative, quantitative, and experimental–that will need to be completed in the next few years. The second workshop provided NSF with a number of action and research recommendations, including a call for additional funding for research on public attitudes toward science.

SSCC in the Considerations of the National Science Board

The products of the two workshops were presented in a pair of reports in November 2010 (Toumey *et al.* 2010) and January 2011 (Guterbock *et al.* 2011). The first report was disseminated for the benefit of the November 2010 workshop, organised and chaired by Thomas Guterbock. Subsequently, Dr Myron Gutmann of NSF presented an oral summary to the National Science Board at its meeting of December 2010 and then he presented a written summary to NSB at its February 2011 meeting. The written report was later incorporated into the minutes of NSB's May 2011 meeting as an Appendix. In a one-page article in *Science* in July 2011 (Bhattacharjee 2011), the Chair of the NSB subcommittee for the *Indicators* affirmed that NSB is "revamping the survey" of public knowledge of science in accordance with the two workshop reports. The 2012 *Indicators* included a sidebar in Chapter Seven, which alerted the reader to the work of the two workshops in 2010, along with changes that will become manifest in the 2014 *Indicators* (NSB 2012). Together these developments indicate that the two workshop reports are being institutionalised for the benefit of future editions of the *Indicators*.

SSCC in Science Media

Another series of developments occurred in other venues. Chris Toumey had a two-page synopsis of the SSCC report published as a commentary in *Nature Nanotechnology* in January 2011 (Toumey 2011). Toumey and graduate research assistant Colin Townsend also disseminated the workshop results in presentations at the Society for Applied Anthropology (Seattle WA, April 2011), the Conference on Science and the Public (Kingston upon Thames UK, July 2011), and at North Carolina State University (Raleigh NC, January 2012).

The article in *Science* from 22 July 2011 (Bhattacharjee 2011) contained three statements that Toumey and Guterbock considered misleading, namely: (1) a statement that the SSCC report intended to "downplay" measures of public knowledge of evolution; (2) an accusation that the members of the National Science Board had been acting from religious motivations when they called for a re-examination of the conceptual framework for public knowledge of science; and (3) that changes were recommended because Americans were not scoring high enough on measures of civic scientific literacy.

In response, Toumey and Guterbock wrote a letter which appeared in *Science* on 7 October 2011 (Toumey and Guterbock 2011). Their letter refuted each of those three misleading statements from the July article. On the first point, the authors quoted from the first workshop report:

> The workshop participants strongly feel that the NSB, the NSF, and the *Indicators* cannot retreat from controversies about important scientific concepts. Evolution is a cornerstone of Biology. Measures and reports of public knowledge of science in the *Indicators* and elsewhere need to explore knowledge of evolution.

Their letter added that the SSCC report recommended expanding measures of knowledge of evolution by including adaptation, natural selection and speciation; also, the topic of evolution should not be limited to human evolution.

> The evolution of plants is germane to questions of genetically modified organisms, for example, and microbial evolution is relevant to our use of antibiotics and vaccines.

In summary, "These recommendations do not downplay evolution, as Miller suggests. Just the opposite–they enhance and expand measures of public knowledge of evolution".

Regarding the second and third statements, Toumey and Guterbock (2011) countered that "We see no evidence that the NSB members were motivated by religious reasons", and, "There is no truth to the allegation that we and our colleagues made those recommendations 'because Americans are not scoring high enough'."

Finally, the report of the October 2010 workshop received a very kind appreciation in the magazine of the British Science Association. Under the title "Knowledge should be of practical value", Anjana Ahuja (2011, 29) noted that "the concept of scientific literacy is changing–and…it desperately needs to change". Ahuja paraphrased the workshop report with these words:

> Instead of measuring how much scientific knowledge nonscientists can muster, and then lamenting the paucity, why not reframe scientific literacy in terms of a need-to-know basis instead of an ought-to-know basis?

The SSCC framework, wrote Ahuja, will "allow people to dig deep into the science that affects their lives". The result, she said, will be that people will make better use of science in their lives: more parents will vaccinate their children; more people will take their entire course of antibiotics; more people will understand an allegation that light bulbs cause cancer; and so on. Ahuja's article was unexpected praise of the workshop report, and the participants could not have asked for a more favourable endorsement.

Conclusion

For the process of measuring and reporting public knowledge of science, the revised conceptual framework described here has a clear and distinct starting point: what kinds of scientific knowledge do people in the public need for purposes of civic engagement with science and science policy, and for purposes of making individual decisions about one's life and one's health, and for purposes of feeding a person's curiosity about science?

Furthermore, the revised framework reveals a series of insights about relations between the public and scientific knowledge: the public is far from homogeneous in its relation to scientific knowledge; public knowledge of science tends to be topical rather than nomothetic; and many persons in the public have a considerable ability to acquire, comprehend and employ scientific knowledge when they need to even if this ability is often underestimated.

The transition from "civic scientific literacy" to "science in the service of citizens and consumers" has a certain value: the audience for the two workshop reports is not a small community of academics unattached from the importance of public knowledge of science. On the contrary, the audience comprises three constituencies: (a) the staff of the National Science Foundation, who are charged with measuring public knowledge of science for the *Indicators*; (b) the members of the National Science Board, who are ultimately responsible for the content of the *Indicators*; and (c) policy makers, researchers and others, who use the *Indicators* as a trustworthy reference work.

And so this framework for putting science in the service of citizens and consumers reflects the ways that non-scientists acquire scientific knowledge and make use of it. This reorientation will shape Chapter Seven of the *Indicators*. If indeed policy makers use the *Indicators* when they think about science, then this new conceptual framework can bring new information and new purposes to discourses about a profoundly important topic: whom does scientific knowledge serve, and how does it serve them?

Notes

1. Participants in the October 2010 Workshop on Public Knowledge of Science (with field of expertise), included: John Besley, University of South Carolina (science communication); Meg Blanchard, North Carolina State University (science education); Mark B. Brown, California State University–Sacramento (science policy); Michael Cobb, North Carolina State University (science policy and survey research); Elaine Howard Ecklund, Rice University (science and culture; religion and science); Margaret Glass, Association of Science & Technology Centers (informal science education); Thomas Guterbock, University of Virginia (survey research methods); A. Eamonn Kelly, George Mason University (learning and cognition); Bruce Lewenstein, Cornell University (science communication); Chris Toumey (organiser and chair of the workshop), University of South Carolina (anthropology of science, especially public scientific controversies). Participants in the November 2010 workshop on implementing the recommendations of the first workshop, included: Nick Allum, University of Sussex (survey methods, public understanding of science); John Besley, University of South Carolina (science communication); Frederick Conrad, University of Michigan (survey methods); Allyson Holbrook, University of Illinois at Chicago (survey methods); Scott Keeter, Pew Research Center (survey methods); Susan Losh, Florida State University (Educational Psychology and Learning); Jeff Mondak, University of Illinois at Urbana-Champaign (political science); Bryce Reeve, University of North Carolina – Chapel Hill (psychometrics, quantitative psychology); David Sikkink, University of Notre Dame (sociology of religion) ; Sally Stares, London School of Economics (social measurement, public perception of science); Roger Tourangeau, University of Michigan and University of Maryland (survey methods); Chris Toumey, University of South Carolina (anthropology of science); Tom Guterbock (organiser and moderator), University of Virginia (survey methods). In addition, the two workshops benefited greatly from the services of graduate research assistants Debbie Rexrode (University of Virginia) and Colin Townsend (University of South Carolina).

References

Ahuja, A. 2011. Knowledge should be of practical value. *People & Science*, September 2011: 29.

Bell, P., B.V. Lewenstein, A. Shouse and M. Feder, eds. 2009. *Learning science in informal environments: people, places, and pursuits*. Washington DC: National Academies Press.

Bhattacharjee, Y. 2010. NSF Board draws flak for dropping evolution from *Indicators*. *Science*, April 9: 150-151.

Bhattacharjee, Y. 2011. New NSF survey tries to separate knowledge and belief. *Science*, July 22: 394.

Bodmer, W. 2010. The BA, the Royal Society and COPUS. *Notes and Records of the Royal Society* 64(Supp. 1): S151-S161.

Culliton, B. 1989. The dismal state of scientific literacy. *Science*, February 3: 600.

Dewey, J. 1981 [1934]. The supreme intellectual obligation. In *John Dewey: the later works volume 9*, ed. J.A. Boydston, A. Sharpe and P. Baysinger, 96-101. Carbondale IL: Southern Illinois University Press.

Guterbock, T., N. Allum, J. Besley, F. Conrad, A. Holbrook, S. Keeter, S. Losh, J. Mondack, B. Reeve, D. Rexrode, D. Sikkink, S. Stares, C. Toumey and R. Tourangeau. 2011. *Measurement and operationalization of the "Science in the service of citizens and consumers" framework.* Report to the US National Science Foundation. http://surveys.virginia.edu/csr/publications/measurement-and-operationalization-science-service-citizens-and-consumers-framework.

Losh, S. 2006. *National Science Foundation surveys of public understanding of science and technology, 1979-2006.* http://www.ropercenter.uconn.edu/data_access/data/datasets/nsf.html#overview.

Miller, J.D. 1983. Scientific literacy: a conceptual and empirical review. *Daedalus* 112(2): 29-48.

—. 1987a. The scientifically illiterate. *American Demographics*, June: 26-31.

—. 1987b. Scientific literacy in the United States. In *Communicating science to the public*, ed. D. Evered and M. O'Connor, 19-37. NY: John Wiley & Sons.

—. 2004. Public understanding of, and attitudes toward, scientific research. *Public Understanding of Science* 13: 273-294.

NAGB (National Assessment Governing Board, United States Department of Education) 2009. *Science framework for the 2009 national assessment of educational progress.* Washington DC: U.S. Dept. of Education.

NSB (National Science Board) 2012. *Science and engineering indicators 2012*. Washington DC: National Science Board.

NSF (National Science Foundation) 2003. *National Science Foundation strategic plan, FY 2003-2008.* Arlington VA: National Science Foundation.

Royal Society 1985. *The public understanding of science*. London: The Royal Society.

Russell, C. 1986. America's scientific illiterates. *Washington Post*, June 2: 1A.

Shen, B. 1975. Science literacy and the public understanding of science. In *Communication of scientific information*, ed. S.B. Day, 44-52. Basel: Karger.

Toumey, C. 2011. Science in the service of citizens and consumers. *Nature Nanotechnology* 6(1): 3-4.

—. 2006. National discourses on democratizing nanotechnology. *Quaderni* 61: 81-101.

Toumey, C. and T. Guterbock. 2011. Justifiable changes to *Indicators* survey. *Science*, October 7: 38-39.

Toumey, C., J. Besley, M. Blanchard, M. Brown, M. Cobb, E.H. Ecklund, M. Glass, T.M. Guterbock, A.E. Kelly and B. Lewenstein. 2010. *Science in the service of citizens and consumers*. Report to the US National Science Foundation.
http://nano.sc.edu/resources/publications.aspx.

Wickson, F., A. Delgado and K. Kjolberg. 2010. Who or what is "the public"? *Nature Nanotechnology* 5(11): 757-758.

Wynne, B. 1992. Public understanding of science research: new horizons or hall of mirrors? *Public Understanding of Science* 1: 37-43.

CHAPTER TWO

NATURAL SCIENCE MEETS SOCIAL SCIENCE: THE NRC'S 2009 REPORT ON "LEARNING SCIENCE IN INFORMAL ENVIRONMENTS"

BRUCE V. LEWENSTEIN

Introduction

One approach in the "public understanding of science" community carries the label "informal science education" (ISE). For example, the United States National Science Foundation's (NSF) funding stream for popular science productions was called until 2012, the "Informal Science Education Program" (now called "Advances in Informal STEM Learning"). ISE focuses on "learning outcomes" for individuals, schools, families, and society. The evidence base that describes informal science, its promise and its effects is informed by a range of disciplines and perspectives, including field-based research, visitor studies, and psychological and anthropological studies of learning. Less influential in the field have been historical or sociological studies of the relationship of science and society.

A 2009 report from the United States National Research Council (NRC) addresses the question of what people actually learn in informal science environments. Because the NRC is the operating branch of the United States National Academy of Sciences[1], its reports must pass a peer review process deeply influenced by the natural sciences and indeed often populated with natural scientists. Thus the report, *Learning science in informal environments* (henceforth, LSIE), might be considered a "scientific" approach to public understanding of science. Although I am a historian of science with a particular interest in public understanding of science–definitely not a scientist or specialist in "learning"–I was somehow deemed sufficiently knowledgeable about relevant issues to be co-chair of the committee that produced the report. In this paper, I will summarise the report, but also call attention to aspects of the report where the social study of science allows us to reflect on what constitutes a "scientific" approach to public understanding of science. We might consider this a form of anthropological study, one revealing "backstage" aspects of the institutional contexts for science in public (Hilgartner 2000).

The "Learning Science in Informal Environments" Report

The NRC regularly produces "consensus reports," which summarise the state of knowledge in particular domains. Although it has frequently addressed formal education, only in 2009, after a four-year process of meetings and consultations, did it release a report on out-of-school learning: *Learning science in informal environments: people, places, and pursuits* (Bell *et al.* 2009).[2] The report was funded by the United States National Science Foundation, which was seeking ways to better target its funding in the area of informal science education. It was a consensus study, which means it was both guided and constrained by the questions: What is the evidence for the findings? To what extent are statements in the literature hypotheses or proposed theory versus conclusions for which the evidence is clear? The NRC's consensus study process draws on a diverse group of researchers to produce an extended literature review. The reports themselves endure a multi-stage peer review process before they receive the NRC imprimatur.[3] The committee discussions are full of debates: What do we know? What do we not know?

From the participant-observer position, I can also say that the consensus process means that the report is a committee-written document, with all the strengths and weaknesses of that genre. It is an extremely valuable report, but there are also aspects of it that may not accord with the findings and traditions of some research communities. It may not meet the needs or interests of specific groups, or may not go as far as some researchers might be willing to push.

The LSIE study builds on two earlier NRC studies, which looked primarily at schools (that is, formal education) and learning. These reports were on *How people learn* (Bransford *et al.* 2000) and *Taking science to school* (Duschl *et al.* 2007). The new report had the goal of tying together and complementing some of the earlier research in the formal education realm and extending it to the informal education community. The 2000 report on *How people learn* was especially influential in the "learning sciences" community (Sawyer 2006). The new report, by focusing outside the formal schooling system, was sometimes facetiously called "How people also learn". More importantly, within the ISE community there was some pressure to raise the profile of out-of-school learning, to make an argument for why more resources in the field would be valuable.

While the immediate impetus for and clear origins of the LSIE report came from the learning sciences community, those of us in the science studies community are aware of many other relevant research traditions. Quantitative studies of "science literacy" began at least as early as the 1950s, with significant public and private investment in various kinds of public understanding, outreach and science popularisation activities since that time (Lewenstein 1994). Although the history of science museums extends back centuries, modern "hands-on" science centres emerged in the 1960s; all kinds of museums reflect the creative tensions among scientific research, science education and entertainment (Lewenstein and Allison-Bunnell 2000; Rader and Cain 2008; Rieppel 2012). The Royal Society's (1985) *Public understanding of science* report (informally, the Bodmer report) led to a flourishing of research on many aspects of science and public (Irwin and Wynne 1996; Ziman 1991). Today, the field is full of journals, conferences, special issues and intellectual ferment.[4] Among the complexities generated by this ferment is a wide range of labels: "learning science in informal environments", "science communication", "informal science education", "public understanding of science", "public engagement in science", "public communication of science", "public learning and understanding of science", "popularisation", "popular science", "*vulgarisation*", "*divulgacíon*", etc. A challenge implicitly (but not explicitly) faced by the committee was to locate its work with respect to these labels.

Another point to notice is the funding: the NSF's ISE programme. At the time, that programme was led by Dr David Ucko. Trained as a chemist, Ucko had been a member of the science museum world for many years, with a reputation as a creative institutional leader. He had spent a brief time at the National Academies helping plan the new Koshland Science Museum. At the National Science Foundation, he was shifting the ISE programme from one that primarily funded production to one with more emphasis on evaluation and on production of knowledge–a term I use deliberately for its resonance with other areas of science studies–about the effects of ISE activities.

The committee that produced the report had fourteen members drawn from a wide range of areas. Some focused on research and evaluation of science learning in informal environments, while others were curriculum experts. The committee had several cognitive and developmental psychologists, as well as working natural scientists. Others on the committee were interested in media issues and in after-school and out-of-school issues. Out of fourteen people, eight had extensive experience in museum settings. In addition to the members, the committee had support from NRC staff members with expertise in educational research.

As historians and sociologists, we are attuned to the personal and institutional dynamics that produce communal documents. Notice what is missing from the previous paragraph: the committee included no experts on history of science (except for my own limited expertise–I was primarily on the committee for my knowledge of science and media issues), art, politics, or community organising. Indeed, we had only a single member whose primary affiliation was in the natural sciences and even she was also affiliated with an environmental outreach centre.

The committee's charge emerged from a formal planning meeting in 2005, followed by negotiations between NSF and NRC, and included the committee's own reflections at its first meeting. The final charge was to identify what theoretical perspectives have relevance for understanding learning in informal environments and to collate evidence for learning; then to use that evidence to understand what the features are of effective environments and what commitments are shared between the formal and informal education communities. The charge's questions were:

1. What is the range of theoretical perspectives, assumptions and outcomes that characterise research on informal science? What assumptions, epistemologies, or modes of learning science are shared between the formal and informal science education environments?

2. How do informal science understanding and practice vary in diverse communities?

3. What evidence is there that people who participate in informal science activities learn concepts, ways of thinking, practices, attitudes and aesthetic appreciation in these settings? What kinds of informal learning environments best support the learning of current scientific issues and concerns (e.g., global warming)? What are the organisational, social, and affective features of effective informal science learning environments vis-à-vis a range of learned competencies/outcomes?

4. Are some learning outcomes unique to informal environments? For example, is there evidence that informal learning environments

support the learning of populations who have been poorly served by school science?

5. What is known about the cumulative effects of science learning across time and contexts? How do learners (young, middle-aged, adolescent, older adults) utilise informal science learning opportunities? How do these opportunities influence learners? Are informal learning experiences designed to suit the developmental trajectories of individuals?

6. What information is needed by practitioners in the field? What information is needed by academics seeking to build and enlarge relevant areas of advanced or graduate study? What information is needed by policy makers to affect policies that include informal environments within the scope of education-directed legislation?

7. What are promising directions for future research? Can common frameworks that link the diverse literatures be developed? If so, what would they look like?

I want to call particular attention to the tension between aspects of the first question ("What assumptions, epistemologies, or modes of learning science *are shared between* the formal and informal science education environments?") and the fourth question ("Are some learning outcomes *unique to* informal environments?"). As noted above, though the LSIE study was an extension of prior work on learning, it was also taking place in an environment of constrained resources for all kinds of education. Some of its supporters hoped that one outcome of the study would be pressure to allocate more resources to informal science education. In the ideal world, such resources might be in addition to resources committed to formal education, but in the real world of political constraints and limited funding, the possibility for conflict between the formal and informal institutional worlds was clear.

I cannot overemphasise the degree to which everything in the report depended on finding valid, reliable evidence based in solid research. As the aphorism says, "data" is not the plural of "anecdote". We needed to find systematic, preferably peer-reviewed, research on which to base our findings. However, this is problematic in the case of informal science education. Much of the evidence about learning from media stories, exhibits, zoo visits, etc., appears in the grey literature, in evaluation reports and newsletter commentaries, and the like. While the ISE practitioner community is confident building on information in the grey literature, the "evidence-based" approach of NRC reports required a different standard. The committee worked hard to understand when the available evidence did constitute data, not anecdote, so that we would not lose the value that so much of the grey literature represents. Even when studies appeared in peer-reviewed literature in science studies, it sometimes struggled to meet the standards that would be used by the natural scientists who would be reviewing the committee's work. Much of the work of science studies' scholars on issues of science and the public consists of historical cases or particular sites of inquiry; the committee had to be clear when and how it could generalise from these works to the level of "consensus" required by NRC reports. Most of us were sympathetic to the case study approach, but the consensus concerns sometimes restricted our ability to make strong claims on some issues.

Ultimately, the report drew on more than 1200 publications, coming from many distinct bodies of scholarly literature. In retrospect, we can see that the literature was biased toward publications from the United States, missed some areas (such as reports on media learning that were appearing just as the committee did its work), tried to include the grey literature, and struggled with questions about the stability of preprints, open-access publishing, web-based institutional reports, and so on. For those of us who study the production of reliable knowledge about the natural world, serving on such a committee is a reminder of the difficulty of establishing any claim to universal knowledge.

The committee commissioned a number of subsidiary reports drawing on the detailed knowledge of others outside the committee. These reports covered areas such as indigenous knowledge, assessment, after-school programmes, and the like. The reports are accessible from the committee's website, as are presentations and materials from the public meetings of the committee. The entire report (both in summary and complete forms) can be downloaded for free from the National Academies Press website, a feature now true of all NRC reports.[5]

The report itself began with broad summaries of theory (especially an ecological perspective on learning) and of the state of assessment. A significant segment of the report identified three "venues" in which learning about science takes place:

> *Everyday learning*. People learn while participating in the activities of daily life, such as hobbies, or family walks, or interacting with family while planning meals and other activities, or meeting with friends, or…. The list, of course, is endless.

Designed environments. Some sites are specifically designed to produce or enhance learning. This venue includes science museums, zoos, nature centres, visitor centres–the kinds of sites of special interest to the readers of this paper.

Programmes. In addition to physical sites designed for learning, much planned learning takes place through formal programmes that are not part of the formal school system. These include after-school programmes for school-age children, adult learning and continuing education programmes available in many communities on a wide range of topics, programmes designed for residents of retirement communities or nursing homes, field trips programmes for adults (such as the Elderhostel system), and so on.

The report also addresses two "cross-cutting themes" that appeared in all venues:

Media. Science learning takes place through science shows on televisions, newspaper and magazine articles, science fiction, cyberspace (ranging from the online versions of traditional media through blogs and Twitter feeds), and through electronic games both online and on handheld devices. Many designed environments have media components, such as media interactives on museum floors. Many after-school programmes use media as diversions (for example, on rainy days), and sometimes use programmes or activities delivered through media.

Diversity. Across all venues, issues of diversity affect learning. These include: issues of access that differ across social and economic classes, and across geography (urban vs. rural, for example); issues of intellectual paradigm that vary across communities (such as the differences between indigenous or native groups and the western European and American residents who colonised much of the world, or the differences between modern medicine, folklore, and other medical systems such as acupuncture); and issues of historical relationships with science governed by racial, ethnic, or language differences.

Finally, the report included conclusions and recommendations organised according to learners and learning; informal environments; promoting learning; relationship between informal environments and formal schools; and the development of a common field of research and practice across the informal environment domains. As with any report generated by academic researchers, the report finished with a call for more research.

One of the guiding themes of the report is "life-long and life-wide learning".[6] Formal schooling fills only a small part of the time of our lives and learning takes place not just in school. If one considers the amount of time one spends out of school even during the school years, but then especially afterwards, during a Biblical three-score and ten lifespan, one realises how much space there is for learning throughout life (life-long), as well as across the day (life-wide). A key element of the report was its focus beyond the school system and beyond children. It looked at how adults learn in informal environments, such as when groups of adults stop to discuss a single exhibit on a museum floor, or when individuals alone examine particular exhibits in great detail.

Six Strands of Learning

The centrepiece of the report is an expanded definition of "learning". The definition is intended to move beyond a general statement, to look at different dimensions of learning. The report identifies six "strands" of learning (Bell *et al.* 2009, 41-47). The "strand" metaphor is deliberate; although we can analytically distinguish between the strands, and one might focus activities or exhibits on a particular strand, ultimately all must be woven together to produce learning capable of supporting the load put on it. Put another way: what actually happens in media presentations, community discussions, or in any other informal environments?

According to the six strands, learners who engage with science in informal environments:

> *Strand 1*: Experience excitement, interest, and motivation to learn about phenomena in the natural and physical world.
>
> To be excited about something, to be motivated to explore, is to learn. Although some people think of excitement and motivation as *prior to* learning, both common sense and research show that people learn best when they are interested in the topic. Therefore, the report argues that excitement, interest, and motivation must be considered *part of* learning.
>
> *Strand 2*: Come to generate, understand, remember and use concepts, explanations, arguments, models and facts related to science.
>
> This strand captures the traditional idea of subject competence: there are certain facts and principles that one learns and learns to use.

Strand 3: Manipulate, test, explore, predict, question, observe, and make sense of the natural and physical world.

For many scientists, the idea of learning "the scientific method" is crucial to learning science. Many popular science or science outreach materials also highlight the ability to learn the processes and tools of scientific work and exploration. Notice that the wording of this strand reinforces an asocial understanding of scientific method at odds with the social construction approach used in most contemporary social studies of science.

Strand 4: Reflect on science as a way of knowing; on processes, concepts and institutions of science; and on their own process of learning about phenomena.

Learning researchers have shown that reflecting on what one has been exposed to is a crucial component of turning experience into learning. In the case of science, this reflection needs to look at not only the concepts and processes of science, but also the institutions and organisations that shape the social world in which reliable knowledge about the natural world is produced.

Strand 5: Participate in scientific activities and learning practices with others using scientific language and tools.

For science, learning can happen not just by learning content and method, but also by participating in real science, through programmes such as "citizen science" or "public participation in scientific research".

Strand 6: Think about themselves as science learners and develop an identity as someone who knows about, uses and sometimes contributes to science.

I consider this strand among the most important in understanding science learning. Consider the common experience of many people who work in science or science-related fields: You are at a party and someone asks what you do. You say something about science. The other person responds: "Oh, I just barely passed my high school science class with the lowest possible grade. I am *so* glad I haven't had to deal with science since then." That person does not believe that he or she *can* learn science. She or he doesn't have an identity as a "science learner". They do not understand themselves to be people who can continue to learn science throughout their lives. Part of learning science is learning that one can learn science, acquiring a self-image or self-identity as a science learner.

Most of these strands are not unique to informal environments or to science. Nonetheless, they have nuances that are important to understand, especially for people in the science communication world. Strands 2 to 5 should look familiar to anyone who is interested in learning; they fit both common sense and other definitions of learning (including a definition used in the earlier NRC report on *Taking science to school*). Strands 1 and 6 are less common, but are nonetheless critical for understanding learning. Moreover, while not unique to informal environments, those two strands highlight affordances that are particularly strong in informal environments.

These strands interweave and reinforce each other. They provide a means of understanding how people of different ages build on their experiences to learn science. For practitioners of informal science education, they provide a tool for understanding the relationship between their work with informal audiences and their work with schools (of all levels). A college student can use the resources in a science museum to supplement materials presented in a formal course. A label can highlight that a curator participates actively in a conservation organisation and is an "advocate" for a particular species. That information moves the visitor beyond the idea of scientists as disinterested observers or inquisitors of nature and recognises the social role that they play. That role might involve research, engagement in the process and the institutions of science, and commitment to public outreach.

I want to call particular attention to the presence of "institutions" in the fourth strand. As a member of the science studies community, I believe deeply that better public understanding of science should include an understanding of the complex social processes that lead to the production of reliable knowledge about the natural world. When I say that people should know something about "the scientific method", I do not mean the "hypothetico-deductive method" or "falsification", the types of methods that might be implied by the third strand. I mean the messy, human, institution- and funding-influenced interaction of actors, networks, social groups, boundary objects, technological momentum, and [*insert here your favourite concept from contemporary history and sociology of science*]. Yet, in the context of a consensus report that had to pass muster with the National Academy of Sciences review process, in which many people on the "science" side of the 1990s "Science Wars" might participate, it was very difficult to defend with "scientific evidence" the insights into knowledge production that historians and sociologists of science take for granted. I am glad that the word "institutions" provides a hook for including the lessons of science studies, but I am also aware of the challenges still ahead for bringing science studies into the mainstream of what it means to "learn" science.

Nonetheless, progress is possible. For example, science museums are especially well-placed to address the social and institutional aspects of science learning identified in strand 4. Museums can offer personal interaction, especially with staff scientists or docents who have personal experience. Science museums can also reflect on their own institutional presence. Some of the largest science museums are major cultural institutions in their cities and countries. Museums such as the Deutsches Museum in Munich, the Science Museum in London, the Chicago Museum of Science and Industry, the Franklin Institute in Philadelphia, the Natural History museums of Vienna, London, New York, and Washington, DC–they are all in substantial, "monumental" buildings. In and of itself, that observation can be an opportunity for reflecting on the place of science in society. Sometimes museum staff may be hesitant to exploit this opportunity, understanding the now-discredited imperial impulses behind some of these monumental constructions (Sheets-Pyenson 1988). But museums can claim the opportunity as a classic "teachable moment", becoming ideal places to explore the relationship between national history and cultures and our knowledge of the natural world both at home and abroad. Science museums can show how science and technology are enmeshed in society by highlighting the funding that comes from wars, imperial exploration, economic needs, etc. They can take advantage of controversies to show the interaction of cultural values, scientific findings and the meanings that populations make of science.

Conclusion

The report includes twenty-five conclusions and recommendations, more than can be summarised here. In part because of this complexity, the National Research Council also produced a companion volume, *Surrounded by science* (Fenichel and Schweingruber 2010), directly targeted to the practitioner community and containing many examples and stories of how to put the findings into practice. Few if any of the examples highlight the social context most important to historians and sociologists of science.

From the experience of serving on the committee that produced the LSIE report, I take several lessons. First, standards of evidence remain a crucial difference in the arguments among scientists, social scientists, humanists, and practitioners. While such a statement approaches a "truism", it nonetheless remains fundamental to understanding how this field of "science and public" can progress. Unlike many areas of intellectual inquiry, this field draws explicitly both intellectually and materially from very different fields, and fostering continuing constructive discussions among these groups requires active work.

The very success of the report–it has been widely discussed in the ISE community and references to it are frequent in grant proposals, requests for proposals and similar documents–leads to the second, related conclusion: despite the difficulty of crossing boundaries (because of language, evidentiary standards, etc.), the field of science and public can bring together ideas from the many fields that comprise it. While people in the field have diverse motivations–improving science literacy, questioning the very notion of science literacy, addressing science policy, understanding public culture, etc.–they have much to learn from each other.

Finally, I think the conversations among the committee members themselves and continuing conversations I have had with members of the committee in the several years since the report was issued are examples of the kind of invisible college, collegial network, virtual team, or other names for collective forms of knowledge production in the field of science and public that need to be encouraged and sustained.

Notes

1. More completely, the NRC functions as the operating arm of "the National Academies" (plural), which includes the National Academy of Sciences, the National Academy of Engineering, and the Institute of Medicine.
2. Because the National Academies and the NRC are quasi-governmental organisations, their reports have semi-official status. Therefore, a disclaimer: what I write here is necessarily an unofficial summary. For the full details, the report can be downloaded at no charge from the National Academies Press website, http://www.nap.edu/catalog.php?record_id=12190
3. For more information about the National Academies' study process, see http://www.nationalacademies.org/studyprocess/index.html
4. One conference, of course, is the "Science and the Public" series that generated this volume. Other recent compilations include the U.S. National Academy of Sciences Sackler Symposium on "The Science of Science Communication" (videos are available at http://www.nasonline.org/programs/sackler-colloquia /completed_colloquia/science-communication.html) and a special issue of *Science and Engineering Ethics* on public engagement (Fisher 2011).

5. The committee's website is
http://www7.nationalacademies.org/bose/learning%20science%20in%20informal%20environment.html.
6. A diagram describing the concept is at
http://life-slc.org/about/citationdetails.html.

Acknowledgements

My thanks to my fellow members on the NRC's "Learning science in informal environments" committee for all their hard work and insight. Some of the purely descriptive language about the LSIE report appears in other documents I have authored or co-authored, including (Shouse *et al.* 2010) and (Lewenstein 2013).

References

Bell, Philip, Bruce V. Lewenstein, Andrew Shouse and Michael Feder, eds. 2009. *Learning science in informal environments: people, places, and pursuits*. Washington, DC: National Academies Press.

Bransford, John, Ann L. Brown and Rodney R. Cocking, eds. 2000. *How people learn: brain, mind, experience, and school.* Expanded edition. Washington, DC: National Academies Press.

Duschl, Richard A., Heidi A. Schweingruber and Andrew W. Shouse, eds. 2007. *Taking science to school: learning and teaching science in grades K-8*. Washington, DC: National Academies Press.

Fenichel, Marilyn, and Heidi A. Schweingruber. 2010. *Surrounded by science: learning science in informal environments*. Washington, DC: National Academies Press.

Fisher, Erik. 2011. Editorial overview. *Science and Engineering Ethics* 17(4): 607-620.

Hilgartner, Stephen. 2000. *Science on stage: expert advice as public drama*. Stanford, Calif.: Stanford University Press.

Irwin, Alan, and Brian Wynne, eds. 1996. *Misunderstanding science? The public reconstruction of science and technology*. Cambridge: Cambridge University Press.

Lewenstein, Bruce V. 1994. A survey of public communication of science and technology activities in the United States. In *When science becomes culture*, ed. B. Schiele, 119-178. Boucherville, Quebec: University of Ottawa Press.

—. 2013. What visitors to science museums can learn about the relation of science and technology. In *Relationships between science and technology as presented in exhibits*, ed. R. Bud and L. Zwisler, in press. Washington, DC: Smithsonian Institution Press.

Lewenstein, Bruce V., and Steven W. Allison-Bunnell. 2000. Creating knowledge in science museums: serving both public and scientific communities. In *Science centers for this century*, ed. B. Schiele and E.H. Koster, 187-208. St. Foy, Quebec: Editions Multimondes.

Rader, Karen A., and Victoria Cain. 2008. From natural history to science: display and the transformation of American museums of science and nature. *Museum and Society* 6(2): 152-171.

Rieppel, Lukas. 2012. Bringing dinosaurs back to life: exhibiting prehistory at the American Museum of Natural History. *Isis* 103(3): 460-490.

Royal Society. 1985. *The public understanding of science*. London: Royal Society.

Sawyer, R. Keith, ed. 2006. *The Cambridge handbook of the learning sciences*. Cambridge, New York: Cambridge University Press.

Sheets-Pyenson, Susan. 1988. *Cathedrals of science: the development of colonial natural history museums during the late nineteenth century*. Kingston, Ont.: Mcgill-Queen's University Press.

Shouse, Andrew, Bruce V. Lewenstein, Philip Bell and Michael Feder. 2010. Crafting museum experiences in light of research on learning: implications of the National Research Council's report on informal science education. *Curator* 53(2): 137-154.

Ziman, John. 1991. Public understanding of science. *Science, Technology, & Human Values* 16(1): 99-105.

Part Two

Public Venues of Knowledge

CHAPTER THREE

PUBLIC UNDERSTANDING OF SCIENCE VIA RESEARCH AREAS IN SCIENCE MUSEUMS: THE EVALUATION OF THE EU PROJECT NANOTOTOUCH

CLAUDIA GEYER, KATRIN NEUBAUER
AND DORIS LEWALTER

Initial Reflections

Nanotechnology is everywhere: in sun lotions, instant meals, clothing, medical treatments, etc. New technologies such as nanotechnology offer great opportunities for a better future, but their use may also involve risks in some cases. Nanotechnology is a rapidly developing technology with a growing importance in people's everyday lives. It affects the public to a great extent. For this reason, it is currently essential for the public to deal with this topic and to have a basic knowledge about the usage of nanotechnology, about the benefits and risks of this new technology, and also about nanotechnological research methods and results. This knowledge is an important prerequisite to be able to deal with this issue in a conscious and reflective manner. However, surveys have shown that when compared to other fields of natural science, such as genetics, the general public has a very low level of knowledge about nanotechnology and only a very vague conception of the associated research approaches (Cobb and Macoubrie 2004; European Commission 2008; Gaskell *et al.* 2006; Kahan *et al.* 2009).

Thus, public understanding of nanotechnology and nanotechnological research, as well as the willingness to deal with such topics, has gained increasing significance in adult education. However, despite numerous efforts and a steadily increasing number of activities, courses and practitioners in science communication in general (Burns *et al.* 2003), and more specifically in the field of nanotechnology (e.g. Alexander *et al.* 2012), there is still a wide gap between scientific research and development in the area of nanotechnology and basic public knowledge of this topic (Zimmer *et al.* 2008).

Based on this fact, an initiative called "NANOTOTOUCH", funded by the European Commission, was established with the intention of creating an innovative place for the general public to learn about nanotechnology and to get in touch with current nanotechnological research methods by watching and talking to real scientists during their work. The present article describes the evaluative research results regarding the cognitive and motivational outcomes of this initiative. In the next two sections, the theoretical background on cognitive and motivational goal variables of the study will be presented. Subsequently, the specific characteristics of the museum as a learning environment and the main ideas behind the NANOTOTOUCH project will be outlined, eventually leading to the objectives of the study. On this basis, the methodological approach of the study and its results are presented and discussed.

Scientific Literacy, Public Understanding of Science and Public Understanding of Research

A central aim of formal and informal education is to impart scientific issues to the public. The construct of scientific literacy has a long tradition. It was originally introduced by Hurd (1958), who defined scientific literacy as an understanding of science and its applications in social experience. Scientific literacy is multidimensional; besides knowledge components and a positive attitude toward science, it also contains components of effective processing of scientific contents (OECD 2007). Two very important and partly similar definitions exist, which have great impact on the design and goals of educational activities in both formal and informal learning environments.

One definition, by Bybee (1997), builds the basis for the PISA assessment of scientific literacy. Based on Bybee *et al.* (2009, 866), a scientifically literate person should be able to: (a) use the knowledge to identify questions, to acquire new knowledge, to explain scientific phenomena, and to draw evidence-based conclusions about science-related issues; (b) understand the characteristic features of science as a form of human knowledge and inquiry; (c) be aware of how science and technology shape our material, intellectual, and cultural environments; and (d) be willing to engage in science-related issues and with the ideas of science as a constructive, concerned and reflective citizen.

Within the context of informal education, the concept of "civic scientific literacy", based on Miller (1998), is rather popular. According to Miller (1998, 2004), a scientifically literate person is able to: (a) understand and respect the impact of science and technology on everyday life; (b) make informed personal decisions on science-related topics; (c) read and understand media reports on scientific evidence; (d) reflect critically on the information included therein; and (e) confidently take part in discussions. Miller (1998) also states that the measurement of civic scientific literacy should include two basic dimensions. The first dimension contains a basic knowledge of the vocabulary of scientific terms and concepts. The second dimension describes an understanding of the processes or methods of science for testing models of reality, thereby focusing on the research regarding a subject matter. Miller also refers to a third dimension, which focuses on the awareness of the impact of science and technology on individuals and on society. However, in his opinion this dimension is difficult to assess due to its cross-cultural differences. The acquisition and transfer of knowledge and information regarding the first two aspects of this broad concept of scientific literacy can take place in various learning environments.

Closely related to the construct of scientific literacy are the concepts of public understanding of science (PUS) und public understanding of research (PUR), which are often used as target dimensions in informal learning environments. Based on Durant *et al.* (1989), PUS describes a dialogue between science and the public aimed at making scientific topics available and understandable for laypersons. PUR focuses on the communication of scientific research (Field and Powell 2001). The acquisition of both of these concepts, as well as the concept of scientific literacy, are closely connected to the public's motivational preconditions for engaging in learning activities regarding science-related topics. As such, in the next section, we focus on relevant characteristics of motivational processes and their prerequisites.

Situational Interest and its Conditioning Factors

As mentioned above, the motivation-related focus of scientific literacy, PUS and PUR depends on the willingness to engage in scientific issues. From the perspective of motivation theory, this quality of motivation is captured by theoretical concepts dealing with self-determined and interest-based motivation (Deci and Ryan 2002; Krapp 2002). Especially in free-choice learning environments, these qualities of motivation play a crucial role, because they describe a central motivational requirement on the part of the audience to get in touch with a topic voluntarily and to be willing to acquire knowledge regarding selected content.

When focusing on interest, one can distinguish between situational interest developed in a concrete (learning) situation and longer lasting individual interest (Krapp 2002). Regarding museum visits, the development of situational interest seems to be a crucial and realistic aim for the broad museum audience and one that is worthy of support. Situational interest describes a motivational quality that occurs in a current (learning) situation and is linked to this situation (Hidi and Renninger 2006; Krapp 2002). Two components of situational interest can be distinguished: catch and hold (Mitchell 1993), which are also called triggered and maintained interest (Hidi and Renninger 2006). The catch component describes the first occurrence of the situational interest. In this phase, a person's attention is drawn to a certain issue, and curiosity toward this issue is aroused. The hold component describes a lasting (stabilised) situational interest in the current learning situation. In this phase, a person wants to deal with a certain issue; he or she perceives the issue as meaningful and wants to learn more about it. The repeated activation of the hold phase is supposed to support the development of an individual's interest in the longer term (Hidi and Renninger 2006; Krapp 2002; Krapp and Prenzel 1992).

From a motivation-theoretical perspective, a relevant condition for the development of interest is the emotional experience during an action (Deci and Ryan 2002; Krapp 2002)–more precisely, the experience of autonomy, competence and social relatedness. According to self-determination theory, these are called *basic needs* (Deci and Ryan 2002), whereby the need for *competence* describes a person's effort to feel "effective". The person wants to master the given demands of a task or learning situation on his/her own, and wants to feel neither unchallenged nor overstrained. The need for *relatedness* states that human beings have a strong wish for satisfactory social contacts; they want to feel accepted and recognised by relevant others. The need for *autonomy* describes a person's effort to be an independent actor. The person wants to set his/her own goals and act in line with his/her integrated self, and does not want to be controlled by others. The extent to which these basic needs are fulfilled is a strong predictor of the development of situational and individual interests. Therefore, it is important to support these experiences in order to foster visitors' interest on both manifestations.

The characteristics and opportunities of the museum context for supporting the development of scientific literacy and interest-based motivation are described and discussed in the next section.

Museums as Places to Support Scientific Literacy and Interest in Science

Science museums are free-choice learning environments (Falk and Dierking 2000) that are highly suitable for communicating scientific topics to a wide audience; as such, they offer the opportunity to support the development of scientific literacy and interest-based motivation. Based on their self-image and their task description (see ICOM 2004), the central aims of museums involve knowledge transfer and fostering visitors' interest development. Museums are unique environments, because they bring an extract of reality to people who usually do not have the chance to see or come into contact with those phenomena (Gibson and Chase 2002). Furthermore, science museums are linked to the scientific community and therefore, they are an ideal platform for communicating current science to a wide audience. They are not only focused on special interest groups or visitors with a basic interest in science, but they also strive to reach the entire range of visitors with all types of attitudes toward science (Overskaug *et al.* 2010; Wersig and Graf 2000). Museums offer a very idiosyncratic range of interactions with science topics. As museums offer many choices and possibilities for creating unique learning activities and as visitors have a heterogeneous background with regard, for example, to prior knowledge and motivation, each visitor has an individual visit experience.

As the learning environment museum is multifaceted and rather different from formal learning environments, Falk and Dierking (2000; Falk 2006) have developed a contextual model of learning that describes and organises the various influencing factors of a museum learning experience. The model involves three overlapping contexts: personal, sociocultural, and physical. It identifies key factors of these three contexts that have a constitutive meaning for the learning process and the course of the visit.

The physical context includes advance organisers and orientation, architecture and design, and reinforcing events and experiences outside the museum. Regarding the physical context, the design of the learning environment (e.g., the degree of structure and activation, tasks or media, non-personal or personal information supply) has proved to be an overall significant factor as learning is highly context-specific (e.g., Bransford *et al.* 1999; Lave and Wenger 1991). Among other things, museum research shows that exhibition design features have an influence on the learning process itself (Bitgood and Patterson 1995; Serrell 1996), as well as on the number of exhibit elements a visitor attends and for how long (Bitgood *et al.* 1994; Serrell 1998). Furthermore, the acquisition of knowledge can be constrained if visitors are not familiar with the museum and, as such, have problems with their orientation in the museum (Anderson and Lucas 1997).

The sociocultural context considers within-group sociocultural mediation and the facilitative mediation by others. As part of the sociocultural context, interaction between the visitors and the museum staff is a crucial aspect, which can be also explained by psychological theories such as those on expert–lay communication (Bromme and Rambow 1998, 2001). It has also been proven in several museum research studies that museum visitors are strongly influenced by the interactions and collaborations they conduct with individuals within their visiting group (Crowley and Callanan 1998; Ellenbogen 2002), as well as outside this group, through interactions with museum explainers, guides and communicators, and even other visitors. Both groups can profoundly influence visitor learning (Crowley and Callanan 1998; Wolins *et al.* 1992).

Relevant factors for the personal context are visitor prerequisites, such as prior knowledge, motivation and expectations, and interests and beliefs. Psychological and museum research have confirmed the influences of prior knowledge and experience on learning in general (Kalyuga *et al.* 2003; Kirschner *et al.* 2006; Kyllonen and Lajoie 2003) and on museum learning in particular (Falk and Adelman 2003; Hein 1998; Roschelle 1995), as it builds the background with which to recognise and select relevant information and integrate it into the individual's body of knowledge. The individual's motivational preconditions and expectations have proven relevant for formal and informal learning in several studies. For example, long-lasting thematic interest that exists prior to the visit has been shown to be highly relevant in the selection of information and exhibits, as well as in the intensity of dealing with the information or exhibits in different learning situations such as the museum context (e.g., Adelman *et al.* 2000; Falk and Adelman 2003; Krapp and Prenzel 2011; Renninger and Hidi 2011). Furthermore, self-efficacy plays an important role in the motivation to deal with new information in an unknown environment, as well as in knowledge acquisition. People must feel able to engage successfully with challenging topics, as this demands cognitive effort. This is why self-efficacy plays an important role in cognitive processing of information, as well as in motivation (Pintrich 1999). People with high self-efficacy might deal with new information more persistently and intensively (Bandura 1977, 1997; Liem *et al.* 2008).

After considering the three contexts of influence described in the contextual model of learning (Falk 2006; Falk and Dierking 2000), we now focus on the specific design of the learning environments developed within the NANOTOTOUCH project. Knowledge transfer in the museum context can be either non-personal (by media and exhibits) or personal (by museum educators and guided tour assistants). In the NANOTOTOUCH project, the concept of personal transfer of knowledge was chosen by establishing a direct communication platform between visitors and nanotechnology researchers.

The Idea of NANOTOTOUCH (NTT)

Based on the theoretical background described above, university–museum cooperation was founded to establish research areas in museums and science centres. Most of the research areas have a fully functional nano-research laboratory situated in the public space of the museum, where researchers conduct their work in full public view. Visitors thus receive an insight into the processes and methods of a modern laboratory by observing how doctorate and graduate students obtain data from their instruments and how these are processed or discarded. This insight enables visitors to explore various aspects of the scientists' daily tasks, as well as to discuss openly any research-related topic with them. These research areas enable visitors to get in touch with nanotechnology by observing and talking to real university researchers working on their research projects, such as their dissertation or Master's thesis (see Ecsite 2011).

In this manner, there is an open dialogue between the scientist and the public. Depending on the visitor's interest, the researcher provides information on a selection of the following topics during his/her interaction with the visitor: an introduction into nanosciences and nanotechnologies; topic of his/her research; and personal motives for doing research, as well as goals, visions, problems and daily routines. Objects, videos, nano-demonstrations and nano-products are all incorporated. The scientist gives nano-demonstrations, explains nano-products and initiates conversations with museum visitors. A significant amount of time is allocated for questions and discussion. This is where the main dialogue between the public and the scientists is expected to take place.

The concept of the research area is based on the public display of research and the corresponding interaction of the scientists with a broad public. Thus, it is immediately apparent that the scientists, their communication skills and their instruments play a central role. Here, close cooperation between the university and the science museum/science centre is absolutely essential, with the former providing the scientists and the research topics, and the latter offering space as well as communication and presentation expertise.

For a better illustration of the research areas, the following describes an example of a visit. A visitor group of two people passes the research area; they stop and watch the scientist carrying out his/her research activities. The scientist initiates a conversation and invites the visitors to ask him/her questions about nanotechnology in general and his/her research activities in particular. The conversation between the scientist and the visitors starts. During the conversation, the scientist presents nano-experiments and explains his/her research instruments. The visitors gain insight into the everyday work of a nano-scientist and learn more about the content of nanotechnology in a way that is perfectly adapted to their prior knowledge.

It is obvious that a scientist working in a public environment is the key point of the NTT approach. Such a "public scientist" is confronted with a number of challenges in the fields of science communication, knowledge exchange and work environment. He or she needs general communication skills, as well as competencies in communicating values and personal information. Therefore, in order to prepare the scientists for this challenging task, professional training was provided in the form of two two-day workshops. The first workshop took part before the implementation of the research areas and focused on the content of social issues connected with nanotechnology and on practical examples and exercises to stimulate the communication skills of the participants. The second workshop took part six months after the setup of the research areas and dealt with gathering, analysing, and disseminating the experiences of the young scientists in the various research areas.

As this description shows, the NTT approach strives to provide a focused and personalised insight into nanoscience and present the manifold processes of research, as well as the positive and problematic aspects of new discoveries and new directions for exploration (Hicks 2011). In doing so, it aims to support the understanding of nanotechnology topics and research, as well as the development of situational interest in this topic. The present study focused on evaluating the extent to which these objectives could be achieved.

Objectives of the Study

The main goal of this study was to determine visitors' situational interest in nanotechnology during their visit to an NTT research area, as well as the potential of these research areas to foster visitors' subjectively perceived knowledge about nanotechnology and research in nanotechnology as three core aspects of scientific literacy, PUS and PUR. Furthermore, the study aimed to identify relevant factors influencing increases in visitors' situational interests and their subjectively perceived knowledge during their interaction with the scientists, in order to optimise the learning environment.

Based on this objective, the following research questions provided guidance for the design and data analysis of the study:

1. To what extent could visitors' situational interest be aroused, and their basic needs satisfaction be fulfilled, during a visit to NTT research areas in museums?

2. To what extent is the experience of a visitor's basic needs related to his/her level of situational interest during a visit to NTT research areas in museums?

3. To what extent did visitors perceive an increase in their understanding of nanotechnological topics and research after their visit to NTT research areas?

4. What personal and situational factors are related to visitors' perceived increase of understanding of nanotechnological topics and research during their visit to NTT research areas in museums?

Method

Design of the Study and Instruments Used

The study was conducted in five museums and science centres in four countries: AHHAA Science Center Tartu (Estonia), Deutsches Museum München (Germany), Museo della Scienza et della Technologia Leonardo da Vinci Milan and Città della Scienza Naples (Italy), Universeum Gothenburg (Sweden). In every research area, at least two scientists communicated with the public and were involved in the survey. During the survey period, all visitors who had communicated with the scientists were asked to give an interview based on a standardised interview guideline after their visit to the research area. The interview guideline was constructed in English and translated into the respective national languages

Scales used	Example item	No. of items	Scales	Cronbach's-α	Source
Perceived increase of knowledge – nano topics	I think I have received an impression of what is meant by nanotechnology	3	1 Not at all 2 Hardly 3 A little 4 Well 5 Very well	.86	New development based on Miller (1998)
Perceived increase of knowledge – nano research	I think I have received an impression of the actual research content the scientist is conducting	3	1 Not at all 2 Hardly 3 A little 4 Well 5 Very well	.85	New development based on Miller (1998)
Communication skills – general	List of several communication skills to be rated (e.g., using basic, simplified language; not too many technical terms)	9	1 Poor 2 Fair 3 Sometimes good 4 Mostly good 5 Always good	.88	Adapted from Cegala *et al.* (1998) and Takahashi *et al.* (2006)
Communication skills – personal	List of several communication skills about own experiences to be rated (e.g., talks about personal experiences; talks about his own opinion)	3	1 Poor 2 Fair 3 Sometimes good 4 Mostly good 5 Always good	.80	Adapted from Cegala *et al.* (1998) and Takahashi *et al.* (2006)
Self-efficacy as to the understanding of nanotechnology	I believe that I am able to successfully comprehend the basics of nanotechnology	3	1 Not at all true 2 Barely true 3 Moderately true 4 Quite true 5 Exactly true	.83	Adapted from Schwarzer and Schmitz (1999)

Situational interest – Catch	To what extent could parts of the research area attract your attention?	4	1 Not at all 2 Hardly 3 A little 4 Quite 5 Very	.87	Lewalter and Geyer (2009)
Situational interest – Hold	To what extent did you find it important to deal with the subject matter of the research area?	3	1 Not at all 2 Hardly 3 A little 4 Quite 5 Very	.85	Lewalter and Geyer (2009)
Basic needs for autonomy, competence, and relatedness	During the visit I had the feeling that I could understand even difficult subject matter	6	1 Not at all 2 Hardly 3 A little 4 Quite 5 Very	.81	Adapted from IMI (Deci and Ryan 2005)

Table 3.1. Scales used in the study

languages by professional translators. It included scales for cognitive outcome variables (self-estimated knowledge increase of nano-content and nano-research), as well as motivational outcome variables of the visit (situational interest, basic needs satisfaction), and situational and personal determinants (perceived communication skills of the scientist, self-efficacy regarding the understanding of nanotechnology) (see Table 3.1). The acceptance of the research area ("Overall, how would you evaluate this research area?" 1–very poor to 5–excellent), prior knowledge about nanotechnology ("How would you describe your expertise in the area of nanotechnology?" 1–very low to 5–very high), self-estimated duration of the visit, interest in nanotechnology as a whole, job relation to nanotechnology, and socio-demographic data were assessed via single items.

The inquiry was conducted based on a standardised instruction. A systematic random sample selection method was used.

Furthermore, more detailed, semi-structured interviews were conducted with a subsample of visitors at the Deutsches Museum in Munich to obtain a more differentiated insight into the visitors' perceptions of the visit. These interviews were also conducted directly after the visitor had finished his/her dialogue with the scientist. These interviews focused especially on the visitor's perceived learning increase.

Description of the Sample

Data was collected from a total of 522 visitors (minimum age 12 years) who were equally distributed over the locations. 110 visitors were interviewed in Munich, 98 in Gothenbourg, 102 in Milan, and 106 each in Naples and Tartu. In addition, semi-structured interviews were conducted with 15 visitors at the Deutsches Museum.

The 522 visitors showed a relatively well-balanced gender spread: 53.4% were female and 46.4% were male. Visitors were between 12 and 82 years old, for an average of 29.53 (SD=15.76) years. Noticeable is the visitors' high level of education even for museum visitors, who are known for their rather high level of education (e.g., Graf and Noschka-Roos 2009). 41.4% of the interviewees had a university degree.

The self-estimated prior knowledge about nanotechnology was rather low (M=2.14, SD=1.02). Most visitors described their knowledge about nanotechnology as very low (n=157, 30.1%) or low (n=190, 36.4%), while 23.9% (n=125) reported little knowledge and only about 9% stated high (n=29) or very high (n=16) knowledge about nanotechnology before the visit. Only a small number of visitors reported having a job related to nanotechnology (n=54, 10.3%). On the other hand, 39.5% (n=206) of the interviewees indicated a general interest in nanotechnology. Regarding self-efficacy in terms of dealing with the subject of nanotechnology, the interviewees on average estimated their level at 3.5 (SD=.92), indicating they feel confident about their own ability to understand the subject of nanotechnology. In summary, the audience can be characterised by rather low prior knowledge, but a high motivation to deal with the topic of nanotechnology.

General Characterisation of the Visit to the Research Area

Regarding the general situation during the visit, it must be stated that most of the visitors (n=316, 60.5%) were already acquainted with the museum due to former visits. Only 38.9% (n=203) were visiting the museum for the first time.

The average perceived duration of the visit was a little over half an hour (M=36.75 minutes, SD=35.82), with a range of 1–300 minutes (see Figure 3.1). Most visitors stayed 10 minutes (n=81), 1 hour (n=67), or 5 minutes (n=61). In general, only about one-third of the visits (30.3%) lasted less than 15 minutes (n=158); 21.8% lasted 15–29 minutes (n=114), 15.1% (n=79) of visitors stayed 30–59 minutes, 141 persons (27.01%) visited the research areas 1–2 hours, and 2.9 % (n=15) of the visitors stayed longer than two hours. 25 persons (4.8%) were unable to estimate the duration of their visit to the research area.

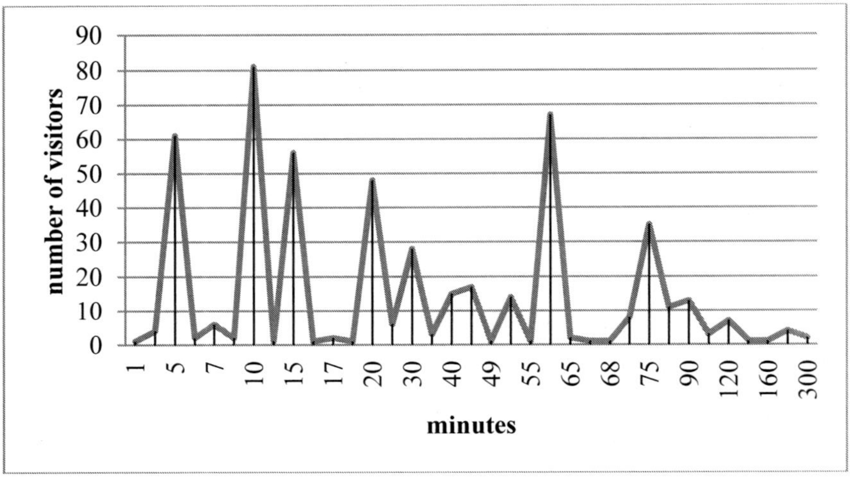

Figure 3.1. Duration of visits in minutes

On average, the visitors were rather satisfied with their visit to the research area and rated the concept as "quite good" (4.20; SD=.72). Very few visitors rated the visit "very poor" (n=7) or "poor" (n=21); 94 visitors estimated the visit as "moderate", while most visitors rated it "quite good" (n=237) or "excellent" (n=140). The scientists' general communication skills were perceived as "mostly well" by the visitors (M=4.21, SD=.72) while the scientists' communication skills regarding their personality was perceived only as "partly well" to "mostly well" (M=3.63, SD=1.07).

Results

Question 1

Besides promoting support for an understanding of nanotechnology, especially nanotechnological research, a central objective of NTT was to foster the public's attitudes and interest in these domains during their visit to a research area, as situational interest is a central prerequisite for dealing with a particular topic in a free-choice learning environment. Therefore, as described above, the motivational variables, situational interest and basic needs were considered in the survey. The findings concerning these variables are summarised in Figure 3.2.

The average level of situational interest "Catch" was slightly higher (M=3.93, SD=.84) than that of situational interest "Hold" (M=3.54, SD=.97). In general, both means of situational interest were quite high. This finding indicates that the scientists managed quite well to arouse the visitors' initial interest and to sustain it during the visit. Regarding "Catch", only 6 visitors had a mean lower than 1.5 and were not "caught" at all, while 17 visitors stated a low level of situational interest "Catch". 87 visitors described a medium level of situational interest "Catch". Most visitors reported a high (n=233) or very high (n=167) level of situational interest "Catch" (missing: n=12).

With respect to "Hold," only 10 visitors could not hold their situational interest at all, 65 visitors stated a low level of situational interest "Hold", and 133 visitors showed a medium level of interest "Hold". Most often, visitors were quite interested (n=189) and 118 visitors even stated a very high level of situational interest "Hold" (missing: n=7).

Based on motivation theory (see above), relevant conditions to support the development of situational interest are experience of autonomy, competence, and social relatedness during an action. Therefore, it is important to promote these experiences to foster the development of visitors' situational interest. Results of the self-reported level of basic needs met during the visit show the following picture. Regarding the experience of competence during the interaction, findings show a medium level (M=3.64, SD=1.08). Furthermore, visitors felt socially related to a high degree with their visitor group (M=4.12, SD=.98). Surprisingly, the social relatedness with the scientist was even higher (M=4.26, SD=.90) and actually reached the highest level of all experiences considered. This means that the visitors felt quite comfortable with the scientist and that the scientist managed to create a pleasant atmosphere during the interaction. With respect to the average experience of autonomy, visitors felt they could actively participate in the conversation (M=3.31, SD=1.28) and that the discussion met their expectations and personal goals quite well (M=3.66, SD=1.12). Furthermore, visitors felt they could participate in the conversation fairly well due to their own initiative (M=3.73, SD=1.30).

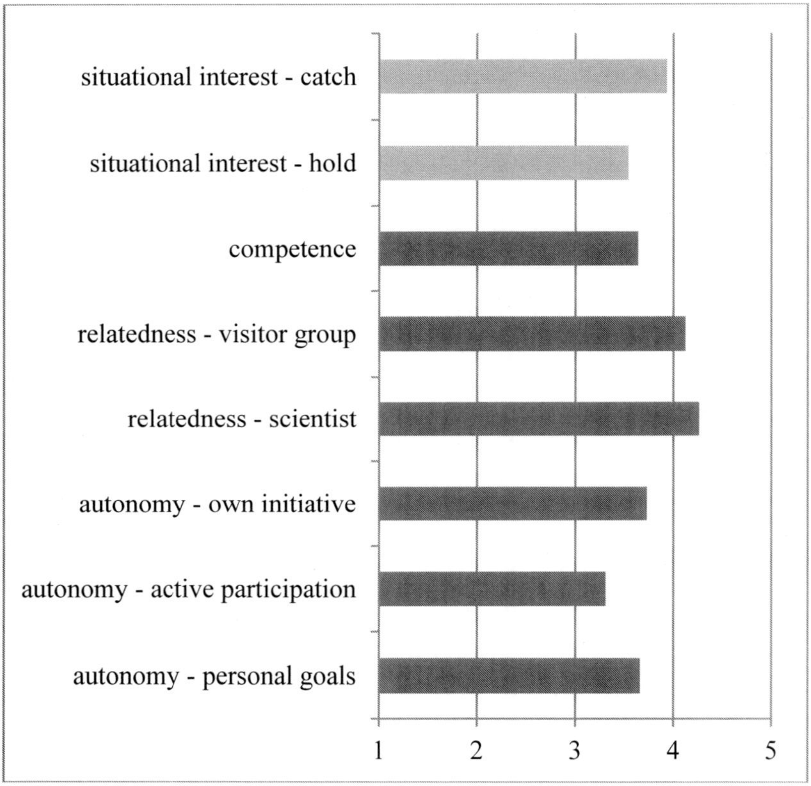

Figure 3.2. Means of situational interest and basic needs satisfaction during the visit

Question 2

In order to optimise the motivational outcomes of the visit, we investigated to what extent basic needs satisfaction was related to the level of situational interest ("Catch" and "Hold"). To answer this research question, multiple regression analysis was conducted with SPSS 19.0. Multiple regression analysis is a statistical procedure that determines the relationship between a dependent variable and several independent variables. Given the independent variables, the conditional expected value of the dependent variable is estimated. A high ß coefficient indicates a high predictive power of an independent variable for the dependent variable, which can be interpreted as a close connection between these two variables (Cohen *et al.* 2003).

Based on the theoretical considerations on motivation mentioned above, the basic needs satisfaction for autonomy (active participation, personal goals and expectations, participation due to own drive), competence, and relatedness (to the visitor group and to the scientist) were taken into account as possible predictive factors.

Situational interest–Catch. Regarding the situational interest "Catch", nearly all variables had a highly significant predictive power, with the exception of the experience of autonomy regarding active participation (Table 3.2). The strongest predictor here was the perceived relatedness to the scientist. In sum, 49% of the variance of the situational interest "Catch" could be explained by the considered variables.

Situational interest–Hold. With respect to the situational interest "Hold", the explained variance by the predictor variables was 42%, which was slightly lower than for the "Catch" component, but still satisfactory. Competence and relatedness to the scientist had the highest predictive power. In addition, autonomy (participation due to the visitor's own drive) and relatedness to the visitor group were significant predictors, while the other two facets of autonomy (expectations and goals; actively participated) were not significantly related to the situational interest "Hold".

	Catch		**Hold**	
	ß	**P**	**ß**	**P**
Competence	.23	**	.30	**
Autonomy – expectations and goals	.13	**	.06	
Autonomy – active participation	-.05		-.04	
Autonomy – due to own initiative	.22	**	.13	*
Relatedness – scientist	.31	**	.32	**
Relatedness – visitor group	.11	**	.09	*
R^2	.49		.43	

* p<.05; **p<.01

Table 3.2. Multiple regression analysis: ß weights and p values of the basic needs on situational interests Catch and Hold

Question 3

A further central aim of the study was to investigate visitors' perceived knowledge increase regarding nanotechnological topics and research.

For a better understanding of the perceived knowledge increase achieved by the visit, we first cite some examples taken from the interviews. In doing so, it becomes more obvious what the visitors understood as perceived knowledge increase regarding nanotechnological topics and nanotechnological research. The first two are taken from interviews about nanotechnological topics and the next two from interviews about nanotechnological research (German statements translated into English):

> The other thing I saw for the first time. There were glass balls, and when you shake them they became bigger and looked like liquid.

> Yes, I especially learned about the manufacturing of nanoparticles.

> Now I've got an idea of the methods, because otherwise I would never have dealt with a grid electron microscope. So now, I know to some extent how to use it and the possibilities you have with it.

> And then you don't know how to investigate everything. Here, you just saw what you can do with it.

Regarding the results of the questionnaire data, we found that, on average, visitors to the research areas rated their knowledge increase regarding nanotechnological content as good (M=3.52, SD=.95). Only 13 visitors answered they had learned nothing new, while 52 answered barely, 155 a little, 174 quite a lot, and 82 a lot. 46 visitors were already familiar with the topic before the visit, and therefore, could not estimate their knowledge increase.

The knowledge increase regarding nanotechnological research was rated somewhat lower—on average, a little (M=3.07, SD=1.04). 35 visitors did not perceive a knowledge increase in this field, 112 persons estimated they barely had a knowledge increase, 175 a little, 120 quite a lot, and 52 visitors stated that they had learned a lot. 28 visitors could not estimate a knowledge increase, as they were already familiar with nanotechnological research before the visit.

Question 4

To identify relevant factors linked to the perceived increase of knowledge of nanotechnological topics and research during the visit to the NTT research areas, multiple regression analysis was conducted again.

Based on the above-described contextual model of learning (Falk and Dierking 2000), personal and situational aspects were considered as possible predictive factors. The following personal factors were taken into account: interest in nanotechnology, job relation to nanotechnology, prior knowledge in the area of nanotechnology, self-efficacy as to the understanding of nanotechnology, and sociodemographic data. Situational factors were in regard to the estimated duration of the visit and both components of perceived communication skills of the scientist. The results are shown in Table 3.3.

Knowledge increase regarding nanotechnological topics. Based on this analysis, the visitor's self-efficacy and the perceived general communication skills of the scientist had the highest predictive power. This indicates that the more self-efficient the visitor was and the more positive he/she estimated the scientist's general communication skills in the field of science, the more he/she estimated to have learned regarding the topic of nanotechnology during the visit. Further significant influencing factors were a job relation to nanotechnology, topic interest, and prior knowledge. Demographic data, such as sex, age and education were not significantly related to the perceived knowledge increase regarding nanotechnological content, nor was familiarity with the museum as assessed through the number of visits to a specific location. Likewise, the estimated duration of the visit and the communication skills with respect to the scientist's own values showed no significant predictive power. In general, 47% of the variance of the perceived knowledge increase regarding nanotechnological content could be explained by the considered factors.

Knowledge increase regarding nanotechnological research. The analysis of knowledge increase regarding nanotechnological research offered slightly different results. Here, both components of perceived scientist's communication skills (general and those regarding own values) were powerful predictors. Regarding personal characteristics, prior knowledge and the visitors' self-efficacy were significantly connected with the dependent variable. Surprisingly, for level of education, a significant negative relation was found. This implies that the higher the visitors' level of education, the lower they perceived their knowledge increase regarding nanotechnological research. In contrast to the perceived knowledge increase regarding nanotechnological topics, no predictive power for topic interest or a job related to nanotechnology was found. Again, neither duration of the visit nor familiarity with the museum were related to perceived knowledge increase regarding nanotechnological research. In this model, the considered variables explained 33% of the variance of perceived knowledge increase regarding nanotechnological research.

In summary, the results indicate that the considered variables contributed to both types of perceived knowledge increase. In both models, the scientists' perceived communication skills were highly connected with the visitors' knowledge increase.

	Perceived increase– nano topics		**Perceived increase– nano research**	
	ß	**P**	**ß**	**P**
Person variables				
Sex	-.03		-.01	
Age	.04		-.03	
Education	-.01		-.19	**
Job relation to nanotechnology	.14	**	.06	
Prior knowledge in nanotechnology	.15	**	.22	**
Topic interest	.14	**	.05	
Self-efficacy	.38	**	.13	**
First visit	-.01		.01	
Characteristics of the situation during the visit				
Scientist's communication–general	.26	**	.17	**
Scientist's communication–personal	.00		.32	**
Duration of the visit	-.04		.06	
R^2	.49		.43	

* p<.05; **p<.01

Table 3.3. Multiple regression analysis: ß weights and p values of person variables and characteristics of the visit on perceived knowledge increase regarding nano-topics and nano-research

Summary and Discussion

The results indicate that the investigated research areas were widely accepted by their audiences.

In particular, the motivational outcomes found were quite good: the concept of the research areas was evaluated as "good" in general; almost nobody rated the visit in a negative way. The basic needs during the visit were satisfied to quite a high extent—the visitors felt quite autonomous, competent, and especially, socially related during the interaction. As expected, a high amount of variance for situational interests "Catch" and "Hold" could be explained through the basic needs satisfaction. Especially noticeable is the very high average regarding the relatedness with the scientists. Thus, the scientists managed to create a pleasant atmosphere during the visit. This is especially relevant because the multiple regression analysis also identified the relatedness to the scientist as one of the strongest predictors for the catch-and-hold phase of situational interest in this learning context. It becomes clear that this variable is a central condition for the development of a situational interest and also for the quality of the dialogue. If a person does not feel comfortable with his/her discussion partner, a conversation hardly starts off and the visitor eventually does not venture to ask any questions that may come to mind. With respect to the situational interest itself, the scientists managed to arouse the visitors' initial situational interest and to sustain it during the visit. This also may be due to the fact that the concept of the NTT research areas is a completely innovative museum design characterised by a high amount of interaction, which in part is highly individualised. The results correspond with those of a number of other studies (in sum, DeWitt and Storksdieck 2008), demonstrating that interactivity and innovation are very popular with museum visitors.

Likewise, the perception of knowledge increase was rated as rather high, especially the perceived knowledge increase regarding nanotechnological topics. Perceived knowledge increase regarding nanotechnological research was somewhat less distinct. Regarding the estimated duration of the visit, this is a substantial knowledge acquisition, although we only assessed the subjective perception of the visitors and therefore have a rather vague indicator of visit duration. The comparatively lower perceived increase regarding nanotechnological research may be due to the fact that knowledge acquisition of the research process requires basic knowledge of the subject matter, which has to be established first, as in most cases the visitors had a rather low level of prior knowledge of nanotechnology. The time most visitors spent in the NTT research areas seems sufficient only to explain the basic content principles, not leaving enough time to explain aspects of nanotechnological research processes.

A further objective of the study was to identify factors connected with the visitors' perceived knowledge increase, in order to optimise the NTT research areas further. Here, the most powerful predictor variable for perceived knowledge increase was the perceived communication skills of the scientists. The better the visitors perceived the scientist's communication skills, the more they felt they had learned. This may be due to a better fit between the visitor's needs and the information provided. Therefore, the combination of the NTT research areas for the visitors on the one hand and the professional training of the scientists on the other have proved to be of value. Professional training in the field of science communication seems to be indispensable for the success of such a concept.

In summary, these findings have to be interpreted carefully as they are based on a rather small sample and the data were gathered in five locations in four different countries. Nevertheless, the results indicate a relatively successful implementation of this new communication format. The research areas seem to be promising new museum concepts that support the understanding of nanotechnology and nanotechnological research and stimulate the development of situational interest in this field. However, further research is needed to confirm these first results regarding these new approaches in science communication in the museum context.

References

Adelman, Leslie M., John H. Falk and Sylvia James. 2000. Assessing the National Aquarium in Baltimore's impact on visitor's conservation knowledge, attitudes and behaviors. *Curator* 43: 33-62.

Alexander, Jane M., Gina Svarovsky, Juli Goss, Liz Rosino, Leight A. Mesiti, Jenna LeComte-Hinely and Christine Reich. 2012. *Study of communication in the nanoscale informal science education network.* Boston: Museum of Science.

Anderson, David, and Keith B. Lucas. 1997. The effectiveness of orienting students to the physical features of a science museum prior to visitation. *Research in Science Education* 27: 485-494.

Bandura, Albert. 1977. *Social learning theory*. Englewood Cliffs, NJ: Prentice Hall.

—. 1997. *Self-efficacy: the exercise of control.* New York: Freeman Press.

Bitgood, Steve, and Don Patterson. 1995. Principles of exhibit design. *Visitor Behavior* 2: 4-6.

Bitgood, Steve, Beverly Serrell and Don Thompson. 1994. The impact of informal education on visitors to museums. In *Informal science learning*, eds. Valerie Crane, Heather Nicholson, Milton Chen and Steve Bitgood, 61-106. Dedham, MA: Research Communications Ltd.

Bransford, John D., Ann L. Brown and Rodney R. Cocking. 1999. *How people learn: brain, mind, experience, and school.* Washington, D.C.: National Academic Press.

Bromme, Rainer, and Riklew Rambow. 1998. Die Verständigung zwischen Experten und Laien: Das Beispiel Architektur. In *Expertenwissen: soziologische, psychologische und pädagogische Perspektiven*, ed. Wolfgang K. Schulz, 49-65. Opladen: Leske and Budrich.

—. 2001. Experten-Laien-Kommunikation als Gegenstand der Expertiseforschung: Für eine Erweiterung des psychologischen Bildes vom Experten. In *Psychologie 2000. Bericht über den 42. Kongress der Deutschen Gesellschaft für Psychologie in Jena 2000*, eds. Rainer K. Silbereisen and Matthias Reitzle, 541-550. Lengerich: Pabst Science Publisher.

Burns, Terry W., John D. O'Connor and Susan M. Stocklmayer. 2003. Science communication: a contemporary definition. *Public Understanding of Science* 12: 183-202.

Bybee, Rodger W. 1997. *Achieving scientific literacy: from purposes to practices*. Portsmouth, NH: Heinemann.

Bybee, Rodger W., Barry McCrae and Robert Laurie. 2009. PISA 2006: an assessment of scientific literacy. *Journal of Research in Science Teaching* 46: 865-883.

Cegala, Donald J., Mary T. Coleman and Jeanine W. Turner. 1998. The development and partial assessment of the Medical Communication Competence Scale. *Health Communication* 10: 261-288.

Cobb, Michael D., and Jane Macoubrie. 2004. Public perceptions about nanotechnology: risks, benefits, and trust. *Journal of Nanoparticle Research* 6: 395-405.

Cohen, Jacob, Patricia Cohen, Stephen G. West and Leona S. Aiken. 2003. *Applied multiple regression/correlation analysis for the behavioral sciences.* Hillsdale, NJ: Lawrence Erlbaum Associates.

Crowley, Kevin, and Maureen A. Callanan. 1998. Identifying and supporting shared scientific reasoning in parent-child interactions. *Journal of Museum Education* 23: 12-17.

Deci, Edward L., and Richard M. Ryan. 2002. Overview of self-determination theory: an organismic dialectic perspective. In *Handbook of self-determination research*, eds. Edward L. Deci and Richard M. Ryan, 3-33. Rochester: Rochester University Press.

—. 2005. *Intrinsic Motivation Inventory (IMI).* Rochester: University of Rochester. http://www.psych.rochester.edu/SDT/measures/intrins.html.

Durant, John R., Geoffrey A. Evans and Geoffrey P. Thomas. 1989. The public understanding of science. *Nature* 340: 11-14.

DeWitt, Jennifer, and Martin A. Storksdieck. 2008. Short review of school field trips: key findings from the past and implications for the future. *Visitor Studies* 11: 181-197.

Ecsite. 2011. *Outcome NANOTOTOUCH project: carrying out live nano research across European science centres and museums.* Brussels: Ecsite.

Ellenbogen, Kirsten M. 2002. Museums in family life: an ethnographic case study. In *Learning conversations in museums*, eds. Gaea Leinhardt, Kevin Crowley and Karen Knutson, 81-101. Mahwah, NJ: Erlbaum.

European Commission. 2008. *Young people and science (Flash Eurobarometer 239).* Luxembourg: Office for Official Publications of the European Communities.

Falk, John H. 2006. The impact of visit motivation on learning: using identity as a construct to understand the visitor experience. *Curator* 49: 151-166.

Falk, John H., and Leslie M. Adelman. 2003. Investigating the impact of prior knowledge, experience and interest on aquarium visitor learning. *Journal of Research in Science Teaching* 40: 163-176.

Falk, John H., and Lynn D. Dierking. 2000. *Learning from museums: visitor experiences and the making of meaning.* Walnut Creek, CA: AltaMira Press.

Field, Hyman, and Patricia Powell. 2001. Public understanding of science versus public understanding of research. *Public Understanding of Science* 10: 421-426.

Gaskell, George, Sally Stares, Agnes Allansdottir, Nick Allum, Christina Corchero and Jonathan Jackson. 2006. *Europeans and biotechnology in 2005: patterns and trends. (Eurobarometer 64.3).* Luxembourg: Office for Official Publications of the European Communities.

Gibson, Helen L., and Christopher Chase. 2002. Longitudinal impact of an inquiry-based science program on middle school students' attitudes toward science. *Science Education* 86: 693-705.

Graf, Bernhard, and Annette Noschka-Roos. 2009. Stichwort: Lernen im Museum oder: Eine Kamerafahrt mit der Besucherforschung. *Zeitschrift für Erziehungswissenschaft. Schwerpunkt: Lernen im Museum* 12: 7-27.

Hein, George E. 1998. *Learning in the museum.* London: Routledge.

Hicks, Paul. 2011. *NANOTOTOUCH: support to outreach and communication in nanotechnology*. München: Deutsches Museum.

Hidi, Suzanne, and Ann K. Renninger. 2006. The four-phase model of interest development. *Educational Psychologist* 41: 111-127.

Hurd, P.D. 1958. Science literacy: its meaning for American schools. *Educational Leadership* 16(1): 13-16.

ICOM. 2004. The definition of the museum. *ICOM News* 57: 4-5.

Kahan, Dan M., Donald Braman, Paul Slovic, John Gastil and Geoffrey L. Cohen. 2009. Cultural cognition of the risks and benefits of nanotechnology. *Nature Nanotechnology* 4: 87-90.

Kalyuga, Slava, Paul Ayres, Paul Chandler and John Sweller. 2003. The expertise reversal effect. *Educational Psychologist* 38: 23-31.

Kirschner, Paul A., John Sweller and Richard E. Clark. 2006. Why minimal guidance during instruction does not work: an analysis of the failure of constructivist, discovery, problem-based, experiential, and inquiry-based teaching. *Educational Psychologist* 41: 75-86.

Krapp, Andreas. 2002. Structural and dynamic aspects of interest development: theoretical considerations from an ontogenetic perspective. *Learning and Instruction* 12: 383-409.

Krapp, Andreas, and Manfred Prenzel. 1992. *Interesse, Lernen, Leistung. Neuere Ansätze einer pädagogisch-psychologischen Interessenforschung.* Münster: Aschendorff.

—. 2011. Research on interest in science: theories, methods, and findings. *International Journal of Science Education* 33: 27-50.

Kyllonen, Patrick C., and Susanne P. Lajoie. 2003. Reassessing aptitude: introduction to a special issue in honor of Richard E. Snow. *Educational Psychologist* 38: 79-83.

Lave, Jean, and Etienne Wenger. 1991. *Situated learning: legitimate peripheral participation*. Cambridge: Cambridge University Press.

Lewalter, Doris, and Claudia Geyer. 2009. Motivationale Aspekte von schulischen Besuchen in naturwissenschaftlich-technischen Museen. *Zeitschrift für Erziehungswissenschaft* 12: 28-44.

Liem, Arief D., Shun Lau and Youyan Nie. 2008. The role of self-efficacy, task value, and achievement goals in predicting learning strategies, task disengagement, peer relationship, and achievement outcome. *Contemporary Educational Psychology* 29: 264-282.

Miller, Jon D. 1998. The measurement of civic scientific literacy. *Public Understanding of Science* 7: 203-223.

—. 2004. Public understanding of, and attitudes toward, scientific research: what we know and what we need to know. *Public Understanding of Science* 13: 273-294.

Mitchell, Matthew. 1993. Situational interest: its multifaceted structure in the secondary school mathematics classroom. *Journal of Educational Psychology* 85: 424-436.

OECD. 2007. *PISA 2006: science competencies for tomorrow's world.* Paris: OECD.

Overskaug, Kristian, Gunnar Holt, Karin G. Hagen, Arild Naess and Morten Steffensen. 2010. An analysis of visitation patterns at the museum of natural history and archaeology, Tronheim, Norway from 1954 to 2006. *Visitor Studies* 13: 107-117.

Pintrich, Paul R. 1999. The role of motivation in promoting and sustaining self-regulated learning. *International Journal of Educational Research* 31: 459-470.

Renninger, Ann K., and Suzanne Hidi. 2011. Revisiting the conceptualization, measurement, and generation of interest. *Educational Psychologist* 46: 168-184.

Roschelle, Jeremy. 1995. Learning in interactive environments: prior knowledge and new experience. In *Public institutions for personal learning*, eds. John H. Falk and Lynn D. Dierking, 37-51. Washington, DC: American Association of Museums.

Schwarzer, Ralf, and Gerdamarie S. Schmitz. 1999. Skala Lehrer-Selbstwirksamkeitserwartung. In *Skalen zur Erfassung von Lehrer- und Schülermerkmalen*, eds. Ralf Schwarzer and Matthias Jerusalem, 60-61. Berlin: Freie Universität Berlin.

Serrel, Beverly. 1996. *Exhibit labels: an interpretive approach.* Walnut Creek, CA: AltaMira Press.

—. 1998. *Paying attention: visitors and museum exhibitions.* Washington, D.C.: American Association of Museums, Technical Information Service.

Takahashi, Megumi, Katsutoshi Tanaka and Hitoshi Miyaoka. 2006. Reliability and validity of communication skills questionnaire (CSQ). *Psychiatry and Clinical Neurosciences* 60: 211-218.

Wersig, Gernot, and Bernhard Graf. 2000. *Virtuelle Museumsbesucher–empirische Studien zur technischen Außenrepräsentanz von Museen.* Berlin: Freie Universität Berlin, Institut für Publizistik- und Kommunikationswissenschaft.

Wolins, Inez S., Nina Jensen and Robyn Ulzheimer. 1992. Children's memories of museum field trips: a qualitative study. *Journal of Museum Education* 17: 17-27.

Zimmer, René, Rolf Hertel and Gaby-Fleur Böl. 2008. *Risikowahrnehmung beim Thema Nanotechnologie–Analyse der Medienberichterstattung*. Berlin: Bundesinstitut für Risikobewertung.

CHAPTER FOUR

FOLLOWING THE STORY OF THE PACEMAKER: FROM "CLEAN ROOM" TO EXHIBITION SPACE

INES HÜLSMANN AND ANA-MARIA RAUS

Introduction

> Opening the door of Room 64 released the hum of activity of four to seven busy people.... They all felt that they were setting the stage for an explosion of knowledge and technology that would profoundly affect the whole practice of medicine. (Bigelow 1984, 943)

When scientists discovered reliable ways to apply the artificial cardiac pacemaker in humans, it was a revolutionary accomplishment for medicine. To tell the history of this device we could go back in time as far as the 18th century (Bing 1992), or focus on the moment in medical history when Earl Bakken, co-founder of the pacemaker company Medtronic in 1949, invented the first wearable cardiac pacemaker.[1] As most new medical devices, it embodied a transfer of technologies, such as the metronome and flashlights, developed for entirely different realms and purposes. Since the first application of the pacemaker, the implementation of the device has extended and nowadays, the pacemaker industry is far away from the initial tinkering which produced it.

Objects within the fields of Science, Technology and Medicine (STM) draw together a variety of different actors from multiple settings in their development, use and evolution. We seek to explore the connections between the spheres of (medical) science, technology and society which the pacemaker captures, in order to uncover overlapping conceptions and meaning-making processes with regards to how science is communicated to the public.

We are choosing the pacemaker as an entry point to the complex and fuzzy space at the intersection of these spheres. By grounding ourselves firmly into the material reality of the technological artefact, we seek to make visible the seamless web of interrelated thoughts and how they turn into conviction and practice. We see the artificial cardiac pacemaker as drawing together different worlds: science, museum and the public. Our approach is a comparison of the relations surrounding the object in two Dutch museums and a Dutch research facility where the pacemaker can be found. Our aim throughout the paper is to follow the object and to find out the different meaning-making processes and ideological exchanges at work when exhibiting and communicating scientific knowledge.

More specifically, we will look at the Boerhaave Museum in Leiden, a traditional science museum; the Continium Science Discovery Center in Kerkrade; and the pacemaker company Medtronic, with scientific research facilities in Maastricht and Heerlen, both located in the South of the Netherlands. These three sites invite for an exploratory study into: (a) the different styles of exhibiting medical devices; (b) the goals and processes stemming from their different approaches; and (c) a further discussion on the possible values of exhibiting scientific knowledge. As we follow the pacemaker throughout these three spaces, questions can be asked as to how and to what means the production and diffusion of knowledge is approached with regards to this object. Medical devices, such as the pacemaker, lend themselves to a deeper understanding. Through telling the story of this medical device, we can think about the ways in which the small pacemaker is transformed from expert knowledge in the laboratory, where the meaning of an object seems obvious, to the museum, where the object's meaning has to be communicated and translated for the public.

To better understand the different levels of meaning that the pacemaker can take across spaces, this research was based on a series of semi-structured interviews[2] with museum curators and scientists, and visits to the science museum, science centre and research facility. The aim of the interviews was targeted at self-perception and self-description of the research participants with regards to their role and perspective on the relation between science, museums and the public.

Furthermore, we ground our paper in social construction of technology studies (SCOTS). There are different theoretical strands in SCOTS. One of them does away with the dichotomy of material objects and human subjects. Together labelled *actants*, both entities affect each other to the extent that they become indistinguishable as parts of an interconnected network. This is the main thesis of Actor Network Theory (ANT), where the object itself is an agent that can ascribe various goals, intentions and possibilities into a system or a network. Relevant for our paper is Latour's (1999) approach to translation processes as taking place between the actants in a network. The relevance stems from our focus on a specific artefact, which we place at the centre of our investigation.

We address three spaces where this medical technology can be encountered, each representing different approaches to working with this device resulting in a variety of translations. The paper's structure follows these three institutions. For each, we look at the place the pacemaker occupies in the respective context, the way the institutions approach science communication, and how the pacemaker is then used as a joining point between certain aims, ambitions and the general public. We then further elaborate on our findings, by comparing the two museums, as well as analysing more in depth the collaboration between the discovery centre and the scientific research facility.

Addressing the world of science museums, we have to be aware that this type of institution is no longer a storehouse of objects but part of the experience economy and seeks to attract visitors (Alexander 2008; Black 1998; Hein 2000; Kirshenblatt-Gimblett 1998). Science museums have an important role in communicating science and in supporting both formal education (outside the classroom) for youngsters as well as life-long learning opportunities for adults (Durant 1992; Hein 1998; Lucas 1983). They must justify the public funding they enjoy by being relevant to the community, increasingly the case since the privatisation movement during the 1990s.

However, they face a number of challenges. Within museums the object takes central stage, but without a certain literacy or adequate ways of presenting it, the meaning of an object often remains hidden and unintelligible (Alberti 2005; Hein 2000; Silverstone 1992). The meanings attached to the devices need to be balanced with the endeavour to construct relevant stories for the museum audience (Cossons 2000; Lindauer 2005).[3] Through different exhibition strategies, museums try to solve their inherent accessibility problem, the fact that they wish to convey specialist knowledge to non-specialist audiences (Boylan 2010). We wish to inquire what kind of background knowledge about science is transported into the exhibition halls (Bauer 2008; Pedretti 2002).

In addition, an increasing body of research is addressing the relations between the public and scientific knowledge–or the public understanding of science (PUS)–with a focus on the challenges and issues at stake when communicating science. Since the Bodmer report in 1985, the field of PUS employs a more nuanced vocabulary to describe processes of science communication: we can now talk of public awareness, public engagement, or science communication. Each term emphasises a different aspect of the public-science relationship. Whatever the label used, it is important to understand the background on which current developments are based (Durant 1999; Rennie *et al.* 2010; van der Sanden *et al.* 2008). We take this study as an opportunity to discuss the matter of public communication of science with scientists to help develop their understanding of the public also (Davies 2008).

We hope our exploratory study, revolving around the artificial cardiac pacemaker, will offer a window into real-life situations when communicating scientific knowledge to the public to reveal hidden translations, challenges and expectations. In the following sections we will address each of the investigated spaces and discuss the pacemaker from the encountered perspectives.

Museum Boerhaave: A Historical Take on the Pacemaker

We start our journey in Leiden at the Museum Boerhaave, which opened its doors as a private museum in 1932 and since 1947, has been the Dutch National Museum of the History of Science and Medicine. Since 1991 it has been located in a historical building with thick yellow walls, which had many different functions throughout the ages–from its original purpose as a nunnery in the 15th century, it was turned into a plague and madhouse, and finally became a university hospital in 1635. Around 1720, the Dutch botanist, physician and humanist Herman Boerhaave (1668-1738) gave his famous sickbed lessons and drew medical students to Leiden from all over the world. It is in honour of his achievements for the natural and especially medical sciences that the museum carries his name.

The museum's collection consists of around 45,000 objects from the beginning of the 16th century to the middle of the 20th century. In general, the acquisition policy prescribes a certain historical distance of about 20 to 30 years and whether a particular item can be considered a key-piece or representative of important developments in the history of science, technology and medicine in the Netherlands. Its permanent exhibition dates back to the 1980s and served as inspiration for the French curators from the Paris Musée des Arts et Métiers on how to exhibit their artefacts. For the last ten to fifteen years, it has been regarded as a prime example of the successful exhibition of scientific and medical technology (Huisman 2010).

We wander quietly through the rooms housing the permanent exhibition. These are relatively closed compared to the more open spaces for temporary exhibitions. Here, we are mostly put in the presence of small-scale objects behind glass and impressive machinery. In a chronological rather than discipline-based arrangement, the Boerhaave exhibits inventions and discoveries from over 500 years of scientific activity. We come across portraits of individual geniuses, as well as a golden armillary sphere from astronomy (16th century), historical maps from geography (17th century), laboratory equipment to experiment with superconductivity from physics (20th century), a chemistry kit (20th century) and Dr Auzoux's anatomical papier-mâché models from medicine (19th century). In general, the exhibition is aesthetically pleasing not least due to the juxtaposition of the monumental atmosphere and high-end glass cabinets, a mixture between dark steel frames and shiny glass, and effective lighting to present the objects almost as artworks.

Finally, we arrive in Room 22, *Man as Machine*, with the main theme of "science put into use for medicine" in the 20th century (Huisman 2010). Here, the pacemaker finds its place in showcase V08, *Revision of Organs*. In the glass cabinet, four pacemakers are presented, distributed over three shelves and surrounded by prostheses for the replacement of soft tissue body parts (such as a heart valve prosthesis, vascular prostheses, breast and penis implants).

The pacemakers differ in shape, material and size. At the very top right corner, there is one round-shaped pacemaker, about 6cm in width, 3cm in depth and 11cm in length, due to two relatively thick plastic electrodes attached to the device. It has a metallic bottom where company name "Cordis" and other details are engraved, while a date seems to be missing. Only its transparent plastic sides allow a glimpse at its insides. Two other pacemakers are relatively similar in their design, round-shaped, about 5 to 6cm in width and length, one being a bit bulkier than the other. Here, their transparent shells are fully visible and we can clearly detect the electrical currents, the generator and five bobbin-shaped batteries. The last of the pacemakers at the bottom in the left corner seems to be from a much more recent date. It has a titanium shell, is significantly smaller and slimmer than the other models. Engraved here is "Cordis, Roden, the Netherlands" and "For Demonstration Only". A small label next to the exhibits indicates "pacemakers" without further indication of their specific application, origins or production dates.

This particular display shows the development of the pacemaker from earlier bulky versions to the small, flat and shiny metallic object we have come to identify as a pacemaker. This arrangement suggests that the scientific practices surrounding the pacemakers have been refined and perfected over the years and that an evolution of the device has taken place. However, explanation is lacking about how exactly the evolution of the device corresponds to the Dutch medical and scientific heritage. Furthermore, keeping in mind the main theme of the room, it is not clear how the device is put into use for medical purposes or the scientific principle behind the technology.

Without further clues to evaluate the artefacts, we read this display as an example of "heterogeneous engineering" (Callon 1995) and the arrangement of the other objects as "allies" which provide the appropriate conditions to understand the object and support a particular message. Here, the larger framework of reference seems to be the augmentation of bodily functions by means of technologies which take over organic deficiencies. What strikes us as strange, however, is that the pacemaker as life-saving device is put together with devices to enhance the quality of life. Moreover, the style of displaying has a disengaging yet commercial feel, as if we as visitors are put in the position of window-shoppers who can simply choose from the technologies while we revise our organs.

For more information about the pacemakers, we used the service on the museum's website which enables online search of the collection (http://www.museumboerhaave.nl/collectie/zoeken/). Although we assumed that we would get to know more about the specific production dates or technical specifications, most of the items are classified in a quarter of a century margin. For the pacemaker this translates to the period from 1975 to 1990. In our perception, it seems rather strange that a museum dedicated to the preservation of past accomplishments would not provide more precise details. Furthermore, our interviews reveal that usually the pedigree of an object is not disclosed to the public. Upon request, the curators mention that the pacemakers enter the collection as donations of outdated models from hospitals or as end-of-range models from companies, preferably Dutch ones.

Mieneke te Hennepe, curator at the Boerhaave, refers to the time in which the concept of the transplant and the replaceable body gained prominence in medical science to explain the place of the pacemaker in a larger historical narrative. According to her, the conglomeration of objects in the glass cabinet is supposed to demonstrate the turn towards organ replacement by technology at the beginning of the 20th century.

However, with the current lack of context about the former "natural habitat" of the objects and their actual usage, te Hennepe questions what the visitor gets from the exhibitions other than "objects in a cocoon" (te Hennepe 2010). Notwithstanding the aesthetic appeal of the exhibition, te Hennepe stresses that an object in itself does not say anything, as you "need something more than just the object. The pedigree of an item tells you if it has an interesting history". Thus, irrespective of their knowledge about the material culture of scientific practice which involves "tinkering" and "trial and error processes" (Huisman 2010), a "social constructivist" stance (te Hennepe 2010) has no room to translate into the way the collection is (currently) presented. By staying faithful to their collection, the overall message remains: scientists advance the progress of society and produce great gifts to society.

Tim Huisman, another curator at the Boerhaave, further emphasises the overall message of the exhibition as suggesting the progress of science. According to him, this has to do with the first generation of curators who were specialists that deeply cherished the aesthetic qualities and cultural significance of the medical devices in the collection. At that time, they wanted to preserve and exhibit these for the "happy few" who had the necessary knowledge and experience to appreciate the items. Huisman distances himself from this elitism and expresses that the aim for the future would be to reach a wider audience and to create long-lasting memories based on personal experiences which connect the visitors to the grand-narrative of Dutch science and culture.

Huisman (2010) relates this to a general shift in orientation and mission:

> up until now we were satisfied with a one-sided traffic of information. We show you the apparatus and you have to look at it and read the signs and that's it. We are now trying to develop ways in which we have more of an exchange, a two-way dialogue.

Huisman also wants to go beyond the celebration of science. Instead of showing science as an autonomous and clean process, his wish is to convey "the handiwork, the problems and the trial and error processes which are behind the great inventions, the great gifts of medical science". What the museum would like to explain is not so much scientific principles as "how science is a cultural phenomenon and how it fits into the context of society". In this way, they present exceptional scientists and their often intimate connections to the objects they used, in order to provide windows into the various scientific cultures. It is in this regard that the technological objects on display can become worthwhile for the larger public.

Within the museum, the guides are regarded as the most important instruments to disclose information about the objects in a meaningful manner as the labels are reduced to a minimum. Emphasised by both interview partners, knowledge about the objects on display can be mediated best through the personal interaction between guide and visitors. Through these tours, the staff members also receive important information about the greatest appeal and shortcomings of the exhibition.

A recent publication by another curator at the Boerhaave, Ad Maas (2010), elucidates how this tension of what can and cannot be shown to the public could be resolved. Here, Maas advocates a move from "showpieces", things appreciated for their sheer aesthetic appeal without context, to "keypieces", things which represent significant developments in the scientific field around which stories can be told. Especially in the context of their independence in 1994, the appeal to a broader audience is important as the Boerhaave now needs to earn 17% of their budget through external funding. In order to be more relevant and attractive to a broader public, they try to move away from their current inaccessible style of exhibiting towards ways to bring their collection to life. While they are satisfied with their temporary exhibitions, they speak almost apologetically about the permanent exhibition, which they expect to be able to change within the next ten years. Here, they want to reach their goal of infusing life into the exhibition through more context and stories about the people behind the instruments, and by employing images and audio-visual media.

As we see, Museum Boerhaave uses the pacemaker as finished product to illustrate the historical development of science. A different approach can be found at our second site of interest, the Continium.

Discovery Center Continium: Pacemaking in the Present

The Continium is a science discovery centre, located in Kerkrade, in the South of the Netherlands. Although it first opened its doors for the public in 2009, the building previously housed the Industrion–Museum for Industry and Society (1998-2008). The Industrion focused on local glass and pottery manufacturers, coal mining practices and cultural history in the South of Limburg. Coal extraction was one of the main industrial activities in the area, starting from the 16th century up until the 1960s. Therefore, mining technologies, personal items and stories related to mining and living conditions of the past constituted the Industrion universe. There, the aim was to immerse the visitor into the past through replicated "authentic" environments. In the Continium, the collection of everyday objects and machines find their place in the *Hall of Fame* section, where reminiscences of the past Industrion continue to bear witness and act as a tribute to the mining legacy, so important for the local community.

Shifting focus, the discovery centre departs from Industrion's mining history towards a more general approach on how science and technology affect people's lives not only in the past, but also present and future (http://www.continium.nl/section.php?sid=content_nl&content_vmd=svw &content_id=about_organisation). Themes such as nanotechnology, sustainability, robotics, or advanced medical technologies feature prominently. These themes are addressed in various ways using the different facilities the discovery centre offers: an exploratory exhibition area, a 3D movie theatre, as well as a "science lab" where visitors can experiment and build their own small scientific projects.

The main exhibition area is open and conducive for exploration. The different themes and exhibits are separated by partial walls, which allows for a fluid transition from one item to the next. A structure on three levels can be observed, meant to guide the visitors: the first identifiable level is an eye-catcher, an impressive device or artefact aimed at attracting the visitors' attention; information panels and the window displays which provide background information on the topic constitute the second level; while the third is represented by touchscreens that offer more in-depth information about the exhibited artefacts. The first layer, the eye-catcher, varies from theme to theme. It can be an anatomic model of the heart, a robot, or a giant rotating tree. Around each of these attention-grabbers, there are circular display areas called stations. Each station groups together interrelated objects and devices focused on a specific theme.

We find the pacemaker in two such stations. The first, the so-called *Heart* station, is also the first visitors can see as they enter the exploration zone. It features an oversized model of the human heart as eye-catcher, where in its proximity the visitors can find a pacemaker display: a plastic casing containing four pacemaking devices–three for the heart and one for the bladder. The engravings on the three heart pacemakers indicate that they are Medtronic products; the origin of the fourth is unknown. One of them is partly open, allowing the visitor to see the circuit boards and wiring inside; another makes visible the batteries needed to power the device; while the third pacemaker is completely enclosed. An adjoining video explains the basic blood circulation and heart functioning mechanisms, and an informational panel describes what a pacemaker is.

With regards to the first station, the pacemaker is presented in an informative fashion. The written details about the device elicit time and patience, especially as the majority of the Continium-goers are children and youth. The visitors do not spend too much time here: there is a lot to read, which is not attractive for the younger visitors, while the adults will dwell more around this display if they have a personal interest in the pacemaker (van Dillen 2010). There is no opportunity to touch the pacemakers or interact with the devices on any level. Nevertheless, the pacemaker is opened up, allowing the visitors to see the inside wires, circuits and batteries. It is the only instance in the settings we visited where the pacemaker as "black box" (Latour 1987) is literally open. In all the other spaces we investigated, the device is closed in its titanium shell, or present as an idea, a symbol for what it can do for the human body. Although the significance of being able to see inside the artefact is debatable, the visitors can learn about the components of the pacemaker or how it functions. However, the device remains distant.

A different situation is at play in the case of the second station, *Me and my body,* which is arranged around three areas of interest: one static and two interactive. The static side is comprised of window displays showcasing prostheses, external heart pumping devices, and pacemaker leads (the connecting wires between the pacemaker and the patient's heart). Although this display seems to offer a history of technological devices that aid human health, including heart-helping devices, there are no actual pacemakers inside. Here, only the leads are presented. The other two areas of interest are a device which mimics CT- and MRI-scanners and a "surgery-room" arrangement. The latter is composed of a human-sized robot in the role of the surgeon, a white mannequin as the patient and a touchscreen that enables the visitor to choose from a range of themes related to technologies which aid and enhance the human body. Two adjoining video screens where further information is offered according to the chosen theme complete the assemblage. Once the visitor chooses the pacemaker from the touch screen, the robot, one of the most loved attractions in the Continium, comes to life and starts to speak about the device. The image of the heart and a connecting pacemaker is then projected on the mannequin and further informative videos are displayed on two neighbouring video screens.

The focus lies here, on how the pacemaker is placed in relation to the human body and what it can do for the human heart. The robot acts as the mediator between the visitors and the pacemaker, a role that the touch screen can be said to have as well. However, although the robot and the touch screen can be considered interactive and attractive for the museum-goers, they represent a veiled way to obtain information. They are merely introducing the technology, but offer no actual "hands-on" learning possibilities (Caulton 1998) with regards to the pacemaker. Still, a learning process might nevertheless take place on the part of the visitors through the overall setting of the "surgery-room", as this display speaks to them on a level they are most likely familiar with–as patients. In this role, facilitated here by the mannequin-patient, the visitors get closer to the device and its importance for human life. This intimacy is enhanced through a personal and bodily experience which the whole arrangement facilitates.

This embodied identification between the visitor and the mannequin-patient can be better understood in terms of translation, as used by Latour (1999, 179, emphasis added), "to mean displacement, drift, invention, mediation, *the creation of a link that did not exist before and that to some degree modifies the original two*". The connection between the device and the heart projected on the mannequin translates into a mental connection between one's own body and the machine–the pacemaker. For the visitors, the pacemaker may become more than a machine, as it connects to the most important organ of the body, the heart. The connection between human organ and technology is strengthened by the ensuing videos displayed on the video screens, explaining how the pacemaker functions inside the body.

Once a connection between visitors and technology has been established, the device can no longer be seen simply by itself, but only in relation to the human body. The meaning of the pacemaker is therefore changed. Instead of just another medical device, it transforms into a technology of vital importance, just as is the human heart.

The translation of competencies, Latour (1999) argues, is fully symmetrical. Applying this symmetry principle to the pacemaker-visitor context, not only does the technological artefact not change its meaning, but the way the visitor thinks of the heart is changed by associating the human organ with the medical object. Following this rationale, the human body is conceived as a machine which can be repaired by the aid of another machine. In other words, between subject and object one can notice a "process of exchanging competences, offering one another new possibilities, new goals and new functions" (Latour 1999, 182). As a result, the pacemaker is not only an innovative and small artefact, but its functions only make sense in relation to the heart; while the heart's meaning transforms to a machine-like organ which can be regulated through the aid of medical technology.

How visitors experience the technologies presented within the Continium is very important for the curators at the discovery centre. In the words of one of its curators, Wouter van Dillen (2010), Continium can be placed in-between traditional "high-end" museums and "science centres which are solely based on playing". In what could best be described as "edutainment"[4], this modern museum tries to engage the visitor in a variety of ways. The three structural levels outlined above, the eye-catcher, the information panels and the touch screens have as the main goal to encourage visitors to engage more actively with the museum's collection.

In doing so, the curators want to showcase advancements in science and technology and aim at grounding the visitors in past and present, but also highlight the possible applications of technoscientific innovations, with an emphasis on future relevance.

The members of the public the Continium mainly addresses as their audience are children and their parents as well as schoolclasses, and as such should be educated about science and technology. Van Dillen (2010) explains that the Continium's strategy is to educate the public: "We want to educate our visitors about past, present, and future, and the role of society, technology, and science, and how those areas interact". The past-present-future outlook is clearly depicted in the discovery centre through the use of objects from the past, present concerns or hot topics, as well as future advancements. Specifically applied to the pacemaker, the temporal layers are visible, too. Older versions of pacemaking devices are displayed, present uses in the form of the artificial cardiac pacemaker and future uses are hinted at in the videos of the surgery-room arrangements.

However, it is not very clear from the exhibition in what way the connection between society, technology and science can be made explicit, because the interaction between the three areas is characterised by mix and overlap. For the pacemaker it is not clear where science stops and technology starts, or how the visitor should read the social factors that contributed to the development of such a device from the display. Recently, awareness has grown on the part of the Continium staff through visitor feedback that a professional guide who offers additional information is needed in order to enhance the meaning of a display. The expert knowledge offers added value and it could provide the missing linkage in the relation between science-technology-society.

A second approach to the relation between the discovery centre and the public is to inspire youngsters in particular to do more science and to become scientists themselves. Van Dillen (2010) stresses the basic message he wants to convey to children visiting the Continium: "you could be that scientist". This goal is reinforced with the presence of the "science lab" in the building, where children especially can experiment with small scientific projects. In spite of this, the pacemaker does not feature in the Continium lab. The *mechanical* workings of the device, although relatively simple in nature are not further addressed via this route. The pacemaker remains enclosed in the various displays, showcased for what it can do rather than how it is made.

A different kind of laboratory becomes important here. Medtronic, a major pacemaker company, has research facilities just ten kilometres away from the Continium, so it is no wonder that all pacemakers exhibited in the discovery centre are branded Medtronic and all the informational videos on the medical device are provided by the company as well. Medtronic scientists worked together with the Continium staff to prepare the pacemaker exhibits and both institutions wanted to create excitement and fascination towards the medical device.

To better understand the relationship between Continium and Medtronic, and to go more in depth with regards to the possible knowledge transfer between the two locations, we will next look at the place the pacemaker occupies in the scientific laboratory at Medtronic and the meanings it receives when used by the scientists there.

The Pacemaker at Medtronic: A Heartfelt Device?

Medtronic is a medical company based in the United States and the world leader in market share in pacemakers. From Medtronic's origins in the field of cardiac rhythm disease management (CRDM) with the pacemaker as main focus, various other medical technologies are currently developed for the treatment of over thirty different chronic illnesses to "alleviate pain, restore health and extend life" (http://www.medtronic.com/about-medtronic/our-mission/index.htm). It founded the Maastricht-based Bakken Research Center (BRC) in 1987 with a focus on scientific research and medical innovations. In close collaboration with medical experts from universities, hospitals and private practitioners, BRC aims at developing life-changing technologies to improve the way chronic diseases are treated.[5]

There are three main groups working at BRC: individual researchers who concentrate on (pre)clinical trials, improvement and alternative uses for the therapeutic devices as well as innovative ideas; engineers who work on the creation of soft- and hardware modification for the devices; and finally the lead researchers who work on better solutions and customisation of the electrodes which connect the devices to the human body.

Moreover, BRC hosts two different kinds of laboratories, one electrical and one mechanical. Within the former, the focus lies on developing, tweaking and perfecting the software for the devices and their electronic components. It is here where they work on the software with which they monitor the performance of the various devices from afar. Moreover, they use the so-called "virtual patient" software to test the performance of the devices. They do so by simulating interferences likely to occur in real-life scenarios and test their hypotheses about pacing variations and algorithm manipulation. The mechanical laboratory is comprised of "machine shops", where parts for prototypes are created and then assembled and tested in the "clean room". The clean room provides the appropriate dust-free environment for the safe fabrication of the devices and the leads.

During our visit to BRC, we see many pacemakers on display as soon as we enter the entrance hall. In three glass cabinets arranged alongside the walls, pacemakers are shown in various guises: pacemakers to support the heart as defibrillator, pacemakers for the bladder, brain and kidney. The display also suggests differences according to the time in which they were created. They are shown in the proximity of other devices produced by Medtronic, making the display a space to showcase their wide range of products. Likewise, in the entrance hall of the European Operations Center in Heerlen, different pacemakers feature among other Medtronic products from different ages, all presented in a large glass cabinet. This showcasing of products for clients and business partners in the more public spaces of the facilities corresponds to common marketing strategies. Interestingly however, as we follow our interview partners through the facilities in Maastricht, the pacemaker is also ubiquitously displayed in the hallways near the researchers' offices and laboratories to which visitors usually do not have access.

Due to the focus on research and innovation at BRC, the pacemaker and its inner workings do not feature predominantly in their work as it is a well-established device. Instead, the emphasis lies more on its applications and future usage, according to the scientists we interviewed (Cornelussen 2010; Receveur 2010). However, the idea underlying the pacemaker, that of targeted electric stimulation of an area of the body for therapeutic reasons is at the core of the scientific thinking of both researchers and engineers at Medtronic. These basic competencies of the cardiac pacemaker are taken to pacemaker-like devices, or from the heart to other organs and later refined through testing processes and clinical trials. The question of "how it actually works" is no longer relevant, as the pacemaker is used as a set of competencies or concept, rather than an actual device.

This, however, does not mean that the cardiac pacemaker as such has become unimportant. Quite the opposite is true judging by the various sites where the pacemaker is on display inside Medtronic facilities. Firstly, showcasing the pacemaker is relevant as part of Medtronic's product portfolio for visitors, business partners and potential clients. Secondly, and more interestingly, the fact that the pacemaker is displayed almost ubiquitously in the more secluded parts of Medtronic sites suggests that the meaning attached to it goes beyond being merely an object. Rather, the pacemaker is proudly showcased as symbol for the foundation on which

Medtronic was built, bears testimony to previous accomplishment and gives hope for future successes. While the pacemaker is by far not the only device Medtronic produces, it still acts as stimulant for thought and seems to set the pace even for the company–or as one researcher expressed: "I cannot sell an idea, I have to have a device" (Cornelussen 2010)–and the pacemaker as most reliable and enduring product is built upon and further developed rather than transgressed. Here, the idea of pacing embodied by the pacemaker becomes an outlook to the world. In this sense, just as with the pacemaker display at the Continium, there is a technological translation happening with regards to this specific technology in the scientific laboratory. The object is translated from a material thing to a symbol, a concept that is then used to further research. The symbol or concept the pacemaker embodies is the idea of pacing through electrical stimulation. The scientists use this idea to translate the concept of the pacemaker to other pacing devices and applications. In this respect, the pacemaker seems to be as much part of their company heritage as it forms part of their present and future.

All our Medtronic interviewees stressed the company's strong commitment to enable and support various activities promoting health in the Dutch Limburg area: through university-related events, open days, local events, conferences, and the media. These activities are aimed at creating awareness of the company, its services and products. In this regard, their collaboration with the Continium stands out.

According to one researcher, science museums and centres in general fulfil many important roles in the communication of science, as "they show you how we progressed" (Cornelussen 2010). He regards it as "very important for people to see we are still progressing, we don't stand still but we are moving forward". As trust is a crucial issue in the health business, he explains that science museums are regarded as reliable sources for information and highly trustworthy for the distribution of scientific knowledge.

He clarifies that while the public has access to all kinds of sources of information, so often these sources cannot be trusted and potential customers are easily misinformed and suspicious due to the exposure to traditional media and the internet. Nevertheless, he is somewhat pessimistic about the transfer of expert knowledge concerning the internal workings of "every very complicated device on display". Instead, science museums should be child-friendly in orientation and could tell "little anecdotes" to get people interested. To him, more important than conveying expert knowledge is to foster a climate of sympathy for science through the celebration of past achievements, so that people get a perspective on the (future) successes of science.

Moreover, while the general public has traditionally trusted the physician's knowledge, patients are increasingly searching for information themselves which can be a burden for the physician. In this regard, science centres informing about the state of the art in current research create awareness, which is important for both physicians and potential customers. In this way science centres and museums are perceived as vital instruments not only to make aware and broaden horizons, but also in shaping the vision of the "bionic person".[6]

While the collaboration between the Continium and Medtronic is in general considered a great opportunity to increase the visibility of Medtronic and promote health, details about the cooperation elucidate a difference in perspective on the aims of communicating science to the public. For Tim Jongen, European Product Specialist for Therapy Management at Medtronic and one of the scientists who collaborated directly with the Continium, working with the museum was a way to increase awareness and stimulate fascination with technoscientific practices. Next to donations of Medtronic products to furnish the showcases and the promotion videos mentioned above, Medtronic scientists also explained the principles behind the pacemaker to equip the staff at the Continium with expert knowledge. Moreover, they offered ideas about how the pacemaker may best be implemented in the Continium's exhibition.

According to Jongen (2010), one of the issues they encountered was the target group of the Continium: young children and their parents. From his personal observations at the centre, he grew concerned about the playful behaviour of youngsters, their short attention span and the parents' neglect of the information panels. He is apprehensive that instead of *edu*tainment, there would only be *enter*tainment. While he acknowledged the fact that the intricate workings of the pacemaker would be difficult to explain to young visitors at the discovery centre, he tried to come up with ideas to make the exhibits interactive and fun, but nevertheless educational. Jongen wished for interactive exhibits as ways to open up discussion and have a multitude of possible outcomes instead of a predetermined one. His aim was to build up excitement instead of giving away the basic principle, which he thought would lead to mere button-pushing.

One of his ideas for an interactive exhibit was to show the effect of deep-brain stimulation through pace-making technology on people with Parkinson's disease. The movement-related symptoms could have been mimicked by a moving platform and the "tremor" regulated by a lever which increases and decreases the electric stimulation. His aim with this exhibit idea was to let people personally experience the symptoms and become familiar with therapeutic possibilities. As a side-effect, visitors would become more aware of people with Parkinson's disease and develop sympathy for sufferers.

Another idea was to design a basic model of the pacemaker for the collection of Continium's "Science Boxes", the "do-it-yourself" kits with which children can tinker and become familiar with mechanical principles. This could have the positive effect of infusing the children with the belief that the world around them can be understood, which he also perceives as one of the main assets of science centres. Another aspect is that over the years, as Jongen bemoans, the number of people enrolling in science and engineering programmes has been declining steadily and it thus becomes ever more urgent to expose children to the fascinating side of science and technology. Through personal gratifying experiences with science and technology, Jongen hopes to ignite fascination and sees the Continium as a possibility to recruit visitors into the world of science (not least for the local division of Medtronic).

In the Medtronic research facility, personal contact with patients as end-users is scarce and one researcher expressed his reluctance to engage with patients directly. For the communication between scientists and patients, physicians have a vital role as they not only mediate the responses from patients in general but are regarded as a source of information about the actual needs of patients. For direct feedback on the performance of their devices, Medtronic concentrates on remote monitoring and makes accommodation for implant support.

While communication to patients on the individual level may not be sought after, from time to time real patients who underwent life-changing pacemaker surgery get invited to speak to the staff in order to endow the scientists' day-to-day practice with a personal meaning. This is regarded as vital in reminding them of their responsibility and honourable service to society and strengthening the bond among staff members (Jongen 2010).

After looking at the pacemaker in the three individual sites where it features, we continue by addressing the device through comparing the different ways it gets imbued with meaning across the three institutions.

Discussion of Findings

Our focus on the pacemaker reveals important aspects of its own time-space-social trajectory. In our discussion, we will first focus on a comparison between the two museum spaces. Afterwards, we will take a closer look at the collaboration between the science centre and the science research centre. We will complement our discussion with an analysis of the different possible ways in which science currently is and could be communicated by the sites under investigation.

When comparing the Continium in Kerkrade with the Boerhaave Museum in Leiden, further meanings of the pacemaker as a technology become clear. As opposed to the approach of the Boerhaave Museum, where the pacemakers were showcased in the glass cabinet as a representation of the evolution of such devices, this is not the focus for Continium. Early versions of pacemaker-like technology are presented (the external heart pump), but a historical narrative is lacking, the focus being on modern pacemakers and the way the devices work inside the human body. The science museum in Leiden also presents artefacts as a representation of culturally significant Dutch heritage, as we saw above, while in the case of Continium, the look is towards the future and how technology can contribute in that respect. These differences between the two institutions spring from the different messages each wants to convey. For Museum Boerhaave, attention is diverted to the finished product of science and the historical development of devices, while for Continium the role technology can play in everyday life as well as supporting the future is where the emphasis lies.

Nevertheless, there are similarities between the two institutions. In both museums, the pacemaker is an important element in the theme of the human body, but here is where the parallel ends: each museum has a different approach to the device stemming from the different goals and perspectives the institution as a whole takes. In the case of the Boerhaave, history and Dutch cultural heritage are key, while in the Continium the active engagement and future outlook predominate.

Another major difference that colours the perspectives of the two museums is the source of the artefacts in their respective collections. For Boerhaave, the technologies are gifts from private owners, companies, or hospitals. Continium has a more commercial-oriented approach, obtaining the device through sponsorships or business collaboration with companies in the area.

From Continium's perspective, geographical rather than temporal position is key. History is not that important, when the emphasis lies on grounding their exhibition within the local culture of the South of the Netherlands. Akin to Continium's precursor Industrion, the regional cultural and technological heritage is vital for their approach. In this respect, Continium builds on this legacy by collaborating with companies in the area and by using local developments to raise awareness of the implications and applications of technoscience. French museologist George Henri Rivière has termed this type of museum an "ecomuseum", an institution grounded in and serving the local culture, becoming "a tool for the economic, social and political growth and development of the society from which it springs" (Fuller 1992, 328). Although we did not look in depth at the effects of Continium on its environs, we could see the ecomuseum goals represented through the Continium-Medtronic collaboration.

Both institutions aim at grounding their place in the local culture: Medtronic supports and enables various activities promoting health in the Dutch Limburg area; likewise, Continium addresses the rapid changes the community has to adapt to. The exchange of knowledge and ideas between the two institutions can be better understood as a shared curatorial responsibility (Söderqvist 2010) with regards to the pacemaker exhibits. The scientists co-operated with the museum curators in order to present the medical technology accurately and they provided the devices and informational videos used in both Continium stations featuring the pacemaker. The curators worked together with the scientists to present the medical artefacts in an educational and fun way to the audience, with interactive displays and attractive arrangements.

In this respect, the collaboration between the discovery centre and the science laboratory was beneficent for both sides, as it functioned as a "construction of authority" (Wynne 1991). For the scientists, the Continium exhibition reinforced the role of science, the knowledge and expertise of pacemaker developers and medical professionals. For the museum, receiving advice and support from the pacemaker experts instated their own expertise on the subject. Furthermore, the pacemaker company is a vital source of scientific knowledge and technologies as well as funding. For Medtronic, Continium functions as a platform to increase public awareness about the possibilities of the pacemaker and the many applications of the device. Their collaboration, therefore, is advantageous for both sides. Throughout the co-operation, the two institutions shared ideals and goals with regards to their approach to the public and the overall strategy of showcasing medical technology. In doing so, awareness of scientific advancements would be enhanced and the visitors would become familiarised with the pacemaker as a device. Nevertheless, even if the two groups of people worked together on presenting the pacemaker, the pacemaker worked differently for the two institutions.

To analyse this paradox, we will return to the different mediations at play when it comes to the pacemaker discussed earlier for the two spaces.

In the case of Continium we discussed the translation of competencies enabled by the mannequin-patient. That specific arrangement made possible an association between the patient's heart and the pacemaker, which, if taken symmetrically, could also influence the way the heart can be perceived as a machine. At Medtronic, the pacemaker functioned as a symbol rather than a technological device up for discussion and it helped the scientists translate the idea of pacing to other medical devices.

Translation, however, is always coloured by the different mediators involved in the specific context where the translation act takes place. Here is where the issue of translability (Benjamin 1923) comes in to flesh out the translation process: some objects or technologies can be said to have higher translability characteristics than others. As the pacemaker is complex, due to the intricate mechanisms that make the device work, it is hard to say if it is easily translatable. If we talk about translability, there are two issues that we have to keep in mind. One is in how much of the device is translatable and whether the "translator" or mediator needs to have the same knowledge as the producer of the translatable device. Focusing on the latter aspect, in both the "clean room" and the exhibition space we see that the role of the mediator is most important: the (virtual) patient as mediator; physician as mediator; mannequin as mediator; a museum exhibit as mediator; a guide who explains and frames information within the exhibition and the other objects in the glass cabinets which provide the context. And especially because the mediator is so important, he or she should be aware of the many translations at play when dealing with technology, as well as the manner in which the object can and will take different meanings across contexts.

While we followed the pacemaker, we also got to know more about the current situation of the three institutions and their view on the public communication of science. One of our findings is that each institution aims at showing science *in* society instead of science *and* society. This shift in attention highlights the co-dependence of science and society and suggests that the facilitation of a dialogue between the two spheres is more important than a one-way conversation (Cozzens and Gieryn 1990; Ede and Cormack 2012). However, the nature of research or the difference between research and development and application is not openly communicated. Showing the uncertainty in scientific practice would address the misconception of science as autonomous process and could enable a more democratic dialogue with science (Bijker 2003, 2006; Bijker *et al.* 2009; Durant 1999).

Through their reflection on the permanent exhibition, the curators at the Museum Boerhaave indicate their motivation to show elements of scientific practice and the social context of science. Currently however, their display of diagnostic and therapeutic machines as landmarks in the history of science applied to medicine references the practice of certain formal sciences to detach objects from their original habitat in order to study them from a depersonalised vantage point (usually mediated by the very instruments on display). This has the effect that the museum tacitly proposes a view of science as autonomous process in which initial "clumsiness" is transcended and scientists merely perfect techniques to make ever smaller and smarter devices. We therefore acclaim their change in attitude to regard these pieces of technology as nodes around which social practices, political and technological and not least economic developments cluster, and would like to see their knowledge featured more prominently in the exhibition space.[7]

The Continium addresses the impacts of science, technology and medicine in everyday life and provides a perspective on the social implications of scientific development at least to some extent. Rather than conveying basic scientific principles, they aim at familiarisation and stimulation of awareness, and provide an overview of the latest developments in science and research. Here, they wish to stress the malleability of the future and the promotion of active engagement with science. However, currently, they do not manage to convey their idea of a reciprocal relation between science and society.

Each institution comes with its own frame or theory about what an object should do in a particular context and both stress the importance of personal guides, which is supported by their visitor feedback (te Hennepe 2010; van Dillen 2010) and museum studies in general (Hein 1998). While each deserves an evaluation on its own ground, we would like to voice a general critique about the lack of showing the cultural historical and socio-economic dimensions of science, technology and medicine. In particular, we are concerned that both institutions do not question the ongoing trend of medicalising the body or reflect more abstractly on the human condition. Instead, the developments stand as facts and self-evident status quo. In both exhibitions, the display has a strong commercial component and reiterates the domineering role of science and technology. In our interpretation, the visitor is somehow silenced and addressed in the familiar role as patient, not necessarily empowered to interrogate what he or she is looking at.

Furthermore, currently hidden mediations could be addressed more openly. The Boerhaave would come closer to exhibit their level of reflexivity through making evident that the museum does not necessarily document the history of STM in the Netherlands but rather what has been collected from this history. In our view, one approach to state this difference could be through a "historicisation" of their permanent exhibition, showcasing it as an example of the "aestheticisation" of science and technology. This adds a level of reflexivity and would tell a more complex story about science in society to provoke visitors to read not only the displays in terms of progress and national scientific prowess but invite them to get involved with the way the objects are currently staged. In this way, the museum curators could more easily translate their knowledge as social historians into the exhibition and as a consequence would come closer to showcasing not a monologue of the past but foster a "dialogic imagination" (Bakhtin 1981; Gielen 2004; Katriel 1993).

We are aware that museums and science centres face particular economic constraints and cannot readily change their exhibitions or set-up. However, by means of taking the pacemaker as an example, we would like to address some other techniques meant to express intended messages or realise specific effects through a different lay-out of the object. Therefore, we take this as an opportunity to promote four alternative routes for meaningful interpretations of the pacemaker and hint at their added value. While not all of these alternatives apply to both institutions, they could nevertheless serve as inspiration.

Firstly, in our opinion, even if the Boerhaave does not aim to explain the internal workings of the pacemaker, it could frame the context of the object differently as to hint at underlying scientific principles. One way could be to piece it together with a metronome to show Earl Bakken's early inspiration for the wearable pacemaker. This kind of combination hints at an important factor of science-in-the-making: the ability to take a principle from one context to another. Instead, the museum relies on the familiarity of the object rather than explaining the actual usage of the device. Keeping in mind that the Boerhaave is a history museum in the service of Dutch heritage, they could focus more on outlining specific dates or establish a clear link to Dutch heritage. By contrast, the Continium explains what a pacemaker is and showcases its various components and general form. However, we find that the place in which the explanation can be found is too remote from the room featuring the pacemaker. It stands to question whether visitors carry the information over and make the link to the leads on display in the glass cabinet. Moreover, the other more interactive parts of the room with the giant heart are likely to direct attention away from the information panel. Furthermore, better attempts could be made to specify the pacemaker's link to society, science and technology respectively.

Our second recommendation concerns the object in a meaningful relation with other objects. Apart from our idea to place the pacemaker next to the metronome which inspired it, another approach would be to deal more consciously with the specific historical narrative of medicine the pacemaker is part of. In a thematic display, the pacemaker could be arranged with other instances of applying electricity in the life sciences. In our view, juxtaposing the pacemaker with Luigi Galvani's 18th century experiments would offer a suitable perspective on the pacemaker.

Thirdly, both the museum and the science centre could use their reputed authoritative voice to address more openly the commercial side of STM. One way in which the underlying commercial interests in STM could be problematised more explicitly would be to reveal the pedigree, the origin of the artefact, instead of merely stating the name and function of the objects. In this way the curators could be more transparent about the (in)direct ways museums and science centres collaborate with companies. In addition, the labels could reveal price, reliability and longevity or other concerns with which objects could be evaluated on the grounds of technical specifications and medical effectiveness. Through adding this level of interpretation, they would be able to provide a meaningful framework to aid visitors' understanding of the particular technology and the network in which it is embedded. In our view, this would be an instance of disruption of familiarity through which visitors could be empowered to discuss societal implications of the vision of the "bionic person", for example whether health-care of the future will be limited to the lucky few who can afford to prolong their life through the transplantation of expensive medical equipment.

Finally, as a general remark, it was to our surprise that all research participants expressed the view that showing how an object works or discussing underlying scientific principles is not necessarily part of their agenda. Our interview partners reasoned that people do not need to know about the complicated components of devices to use them (Cornelussen 2010) and that those who want to know more will look for information themselves (Huisman 2010). Here, special reference was made to the TV (te Hennepe 2010) and the washing-machine (Jongen 2010). Especially mentioning these two machines strikes us as strange, because of their mechanical nature. Tube television sets and washing machines are among the prime examples where people tinker with technology in households to avoid costly repairs. While we agree that it would be hard to explain (not to mention repair) the pacemaker in its current form with its increasingly digital components, we believe that concealing the share of engineering in making the pacemaker hampers the development of scientific vocations by the science museum, discovery centre and company, respectively. As recruitment was one of the aims of communicating science to the public, they miss a great opportunity to use the pacemaker as means to ignite fascination and speak to the public on grounds which are potentially meaningful.

Conclusion

We set out to research how a technological artefact is used to transfer and communicate knowledge and thereby acquires different functions from its ordinary life-cycle as medical technology. We picked the pacemaker as our object of interest because it has a specific place in the "man-as-machine" narrative which is an interesting and still to be written chapter in the history of medicine. Apart from its relevance in medical research itself, focusing on the pacemaker led us to three different sites where we investigated both "backstage areas" of knowledge-in-the-making and "front stages" of exhibiting ready-made-science.

In terms of the different styles of exhibiting the pacemaker as a medical device, we have come to observe that the object is never experienced in isolation, but is presented together with "allies", such as other objects in showcases, videos and museum arrangements. In addition to the way the object is framed by the host institutions, the allies contribute to the establishment of an environment in which the artefact can be rendered comprehensible for the target audience. Together with these allies, social actors, involved in the preparation of the exhibitions (curators, scientists, guides) further shape the meanings of the device. In the process of presenting the medical technology to the public, the institutions' goals and approaches colour the object. In various ways, they all translate its meaning and enhance our experiences with it. Keeping in mind that these translations are at work, additional reflection on these meaning-making processes and exchange of competencies between the different spheres could represent an added value to the way scientific knowledge is currently exhibited.

Our first level of analysis concerns the pacemaker in the setting of a traditional science museum and reveals some of the challenges such a museum may currently face in communicating science to the public by means of objects. While the Boerhaave Museum in Leiden exhibits the object to signify scientific progress and illustrate the growing sophistication in medical history, it is in a period of transition from this downright celebratory stance towards a translation of the curators' constructivist perspective into a more visitor-focused way of exhibiting.

Human agency, guides in particular, is one of the most important ways in which they try to disclose the meaning of objects. Through storytelling, they try to infuse life into the scientific endeavours of the past. Paradoxically, the museum has the policy of distancing itself from current science-in-the-making and favours the approach of letting meaning grow over time in order to decide what is worth exhibiting and collecting.

Our second level of analysis focuses on the pacemaker exhibits at the science discovery centre and lays bare some aspects of the seamless web in which the meaning of an object is entangled. While the Continium in Kerkrade collaborates with experts in pacemaker research, the ways in which this knowledge finds its way into the exhibition area have to be translated for a specific audience. Here, the visitor is appealed to in his role as patient, where the meaning of the pacemaker is intimately connected to the bodily experience of transplanting a pacemaker. The discovery centre tries to enthuse its visitors about science and to showcase the ways in which everyday life is imbued with scientific knowledge and technologies. Approaching its collection in terms of past, present and future, Continium takes an educational stance towards its audience.

Our third level of analysis looks at the pacemaker in the setting of a medical research facility and reveals the scientists' motivations and challenges in communicating their work to the public. The scientists' ideas about possible ways to present their technologies are targeted towards raising awareness about their devices. This hints at aspects about the mutual dependence between device manufacturing firms and their clients. Moreover, the pacemaker also functions as a symbol for possibility throughout the scientists' work and may also serve to enhance their collective experience. For the promotion of active engagement with science, Medtronic closely collaborated with Continium. Their goals and aims in this transfer of knowledge had to be translated for the exhibition space, whereby aspects of the intended message got concealed or were not (yet) realised.

Overall, looking at the translation processes at work when dealing with the pacemaker in the three institutions, we see that the device can yield different and maybe even unexpected meanings according to its circumstances. It is embedded in a rich web of relations and its comprehension draws on roles we intuitively understand but which are difficult to transfer.

However, when comparing the two museums, we come to understand that this rich network of relations and meaning-making processes could be further extended and utilised for the presentation of the pacemaker to the public. The two institutions could address the source of the artefacts in their respective collections, as well as better emphasise their respective roles: Boerhaave with a strong historical and narratorial stance, and Continium as an ecomuseum in its regional area collaborating with local companies. The discovery centre's collaboration with Medtronic revealed how institutions exhibiting science and institutions creating or developing science can enjoy shared curatorial responsibilities when working together and can each contribute to the other's construction of authority in their respective fields. Nevertheless, a better reflection on this collaboration would only enhance the exhibit itself and would show even more the reciprocal relationship between science and society (science *and* society, rather than science *in* society). This reflection, geared towards addressing openly the hidden mediations and translations at play, would contribute to a more empowered audience who can reflect on the connections between objects in an exhibition, better understand the pedigree of the artefact, or question the commercial aspects of medical technologies. All these aspects could lead to an increased fascination and understanding of the technoscientific objects that these institutions present and employ by a more open and dialogic route for the benefit of all sides involved: science, museums and public.

Notes

1. Bakken, student of electrical engineering at the University of Minnesota, had a garage-based company for customising and repairing medical devices. In 1957, he was approached by Dr C. Walton Lillehei, a pioneering heart surgeon, who lost one of his young heart patients due to a power outage. The pacemakers which were used then had to be plugged into wall-outlets to function. Bakken took the revolutionary step of combining several technologies, i.e., transistors and small batteries for the construction of his external pulse generator, an idea he borrowed from a circuit design for a metronome he read about in an electronic magazine for hobbyists (Jeffrey 2001). Basic principles of bioelectricity and handicrafts led to the device which supports our natural pacemaker so as to enable survival. A thorough analysis of the meanings attached to Bakken's first transistorised pacemaker, the Medtronic model "5800" has been done by Rhees and Jeffrey (2003). In the honour of Earl Bakken, the Bakken Museum in Minneapolis exhibits the history and nature of electricity and magnetism in science: http://www.thebakken.org/ (Rhees 2009; Stillings 2001).
2. All interviews and visits were conducted in the period November–December 2010, and therefore all our findings are bound by that specific moment. Later changes in exhibitions, set-up, approaches and conceptions could not be taken into account. Therefore this paper represents a slice in time and space of the state of affairs with regards to the pacemaker in these various spaces in 2010.

3. Medical museums are especially revealing as they initially served as institutions for research, turned then to serve the objectives of medical instruction and health propaganda and are now devoted to present the general public with the spectrum of social and cultural issues surrounding health and healing (Arnold 1996).
4. A term van Dillen did not use during the conversation but was indicated as his function on his name tag: "Wouter van Dillen. Productontwikkeling [product development]–Edutainment".
5. The company manufactures among others heart valves, angioplasty catheters and coronary stents; implantable neurostimulation devices for pain management, sleep apnea and the treatment of essential tremor in Parkinson's disease; neurosurgery products; and implantable drug delivery systems.
6. The "bionic person" refers to the idea of enhancing the body of a human being in whole or in part by electromechanical devices. As a medical company which supplies these technologies, Medtronic envisions the possibilities of building smaller and sophisticated devices to implant more than one.
7. Museum Boerhaave is currently working on a redevelopment of its permanent exhibition. In the new plans, medical history and medical technology will be presented in a significantly different manner addressing patient history, the development of the transparent body and ethical issues among others (information received via email correspondence from the museum curators, November 2012).

References

Alberti, Samuel J.M.M. 2005. Objects and the museum. *Isis* 96(4): 559-571.

Alexander, E.P. 2008. *Museums in motion: an introduction to the history and functions of museums*. Plymouth: Altamira Press.

Arnold, Ken. 1996. Time heals: making history in medical museums. In *Making histories in museums*, ed. Gaynor Kavanagh, 15-29. London: Leicester University Press.

Bakhtin, M.M. 1981. *The dialogic imagination: four essays*, ed. M.J. Holquist. Austin: University of Texas Press.

Bauer, Martin W. 2008. Paradigm change for science communication: commercial science needs a critical public. In *Communicating science in social contexts: new models, new practices*, eds. Donhong Cheng, Michel Claessens, Nicholas Gascoigne, Jenni Metcalfe, Bernard Schiele and Shunke Shi, 7–25. New York: Springer.

Benjamin, W. 1923. The task of the translator. In *Illuminations*, ed. Hannah Arendt. New York: Schocken Books.

Bigelow, W. 1984. The pacemaker story: a cold heart spin-off. *Canadian Medical Association Journal* 131(8): 943–955.

Bijker, W.E. 2003. The need for public intellectuals: a space for STS. *Science, Technology, & Human Values* 28(4): 443–450.

—. 2006. Why and how technology matters. In *Oxford handbook of contexual political analysis*, eds. R.E. Goodin and C. Tilly, 681–706. Oxford: Oxford University Press.

Bijker, W.E., R. Bal and R. Hendriks. 2009. *Paradox of scientific authority: the role of scientific advice in democracies.* Cambridge: MIT Press.

Bing, R.J., ed. 1992. *Cardiology. the evolution of the science and the art.* Chur, Switzerland: Harwood Academic Publishers.

Black, G. 1998. *The engaging museum: developing museums for visitor involvement.* New York: Routledge.

Boylan, Patrick J., ed. 2010. *Running a museum: a practical handbook.* Paris: ICOM

Callon, M. 1995. Four models in the dynamics of science. In *Handbook of science and technology studies*, eds. Sheila Jasanoff, Gerald Markle, James C. Petersen and Trevor Pinch, 29-64. Thousand Oaks: Sage.

Caulton, T. 1998. *Hands-on exhibitions: managing interactive museums and science centres.* London: Routledge.

Cornelussen, Richard. 2010. Personal communication, 29 November.

Cossons, N. 2000. Museums in the new millennium. In *Museums of modern science,* ed. S. Lindqvist, 3-17. Canton, MA: Science History Publications and The Nobel Foundation.

Cozzens, S.E., and T.F. Gieryn, eds. 1990. *Theories of science in society.* Bloomington: Indiana University Press.

Davies, Sarah R. 2008. Constructing communication. Talking to scientists about talking to the public. *Science Communication* 29(4): 413–434.

Durant, John. 1992. Introduction. In *Museums and the public understanding of science*, ed. John Durant, 7–14. London: Science Museum.

—. 1999. Participatory technology assessment and the democratic model of the public understanding of science. *Science and Public Policy* 26(5): 313–319.

Ede, Andrew, and Lesley B. Cormack. 2012. *A history of science in society: from philosophy to utility.* Toronto: University of Toronto Press.

Fuller, Nancy J. 1992. The museum as a vehicle for community empowerment: the Ak-Chin Indian Community Ecomuseum Project. In *Museums and communities: the politics of public culture*, eds. Ivan Karp, Christine Mullen Kreamer and Steven D. Lavine, 327-365. Washington and London: Smithsonian Institution Press.

Gielen, P. 2004. Museumchronotopics: on the representations of the past in museums. *Museum and Society* 2(3): 147–160.

Hein, G.E. 1998. *Learning in the museum.* London: Routledge.
Hein, H.S. 2000. *The museum in transition: a philosophical perspective.* Washington: Smithsonian Institution Press.
Huisman, Tim. 2010. Personal communication, 30 November.
Jeffrey, K. 2001. *Machines in our hearts: the cardiac pacemaker, the implantable defibrillator, and American health care.* Baltimore: John Hopkins University Press.
Jongen, Tim. 2010. Personal communication, 15 December.
Katriel, T. 1993. "Our future is where our past is": studying heritage museums as ideological and performative arenas. *Communication Monographs* 60(1): 69–75.
Kirshenblatt-Gimblett, B. 1998. *Destination culture: tourism, museums and heritage.* Berkeley: University of California Press.
Latour, B. 1987. *Science in action.* Cambridge, MA: Harvard University Press.
—. 1999. *Pandora's hope.* Cambridge, MA: Harvard University Press.
Lindauer, Margaret A. 2005. From salad bars to vivid stories: four game plans for developing "educational successful" exhibitions. *Museum Management and Curatorship* 20(1): 41–55.
Lucas, A.M. 1983. Scientific literacy and informal learning. *Studies in Science Education* 10(1): 1–36.
Maas, A. 2010. The storyteller and the altar: Museum Boerhaave and its objects. In *The exhibition as product and generator of scholarship [preprint 399]*, eds. S. Lehrmann-Braus, C. Sichau and H. Trischler, 59-69. Berlin: Max-Planck Institut für Wissenschaftsgeschichte.
Pedretti, Erminia. 2002. T. Kuhn meets T. Rex: critical conversations and new directions in science centres and science museum. *Studies in Science Education* 37(1): 1–41.
Receveur, Rogier. 2010. Personal communication, 8 December.
Rennie, Leonie J., Susan M. Stocklmayer and John K. Gilbert. 2010. The roles of the formal and informal sectors in the provision of effective science education. *Studies in Science Education* 46(1): 1–44.
Rhees, D. 2009. From Frankenstein to the pacemaker. *IEEE Engineering in Medicine and Biology Magazine* 28, doi:10.1109/MEMB.2009.933571.
Rhees, David, and Kirk Jeffrey. 2003. Earl Bakken's little white box: the complex meanings of a first transistorized pacemaker. In *Exposing Electronics*, ed. Bernhard Finn, 75-115. London: Science Museum.
Silverstone, Roger. 1992. The medium is the museum: on objects and logics in times and spaces. In *Museums and the public understanding of science*, ed. John Durant, 34–43. London: Science Museum.

Stillings, Dennis. 2001. The Bakken: a library and museum of electricity in life. *Journal of Scientific Exploration* 15(2): 255–266.

Söderqvist, Thomas. 2010. The participatory museum and distributed curatorships. *NTM Zeitschrift für Geschichte der Wissenschaften, Technik und Medizin* 18(1): 69–78.

te Hennepe, Mieneke. 2010. Personal communication, 30 November.

van der Sanden, Maarten C.A., and Frans J. Meijman. 2008. Dialogue guides awareness and understanding of science: an essay on different goals of dialogue leading to different science communication approaches. *Public Understanding of Science* 17(1): 89–103.

van Dillen, Wouter. 2010. Personal communication, 1 December.

Wynne, B. 1991. Knowledges in context. *Science, Technology, & Human Values* 16(1): 111-121.

CHAPTER FIVE

THEATRESCIENCE 2002-2013

JEFF TEARE

Context

Theatrescience was formed in 2002 by Rebecca Gould and Jeff Teare. Previously they had run the Welsh new-writing theatre company, Made in Wales, and then worked on the Wellcome Trust's "Science Centrestage" project. Before that, among other things, Jeff had been Associate Director of the Theatre Royal, Stratford East in London.

Theatrescience's main concern is to explore the cultural, political, economic and ethical/moral issues of biomedical science. It uses theatre and drama to illuminate science and it uses science to enrich theatre. In practice this has meant that Theatrescience has introduced theatre practitioners and audiences to subject areas they may not otherwise come across and has introduced a means of communication and discussion to scientists that they generally would not have otherwise considered.

Theatrescience workshops and performance development processes have also got theatre practitioners and scientists to work closely together. Post-show discussions, workshops and other public engagement activities have also involved scientists, medical practitioners, students and the general public in the development of their dramatic work. This has been especially true of some of our work in India (see below) when "scratch" (work-in-progress) performances have been mounted for audiences of research scientists whose comments have then been incorporated into the final script. All of our scripts go through a workshop phase when they are open to discussion and most receive a public reading before production to begin to gauge possible reaction. Post-show discussions (formal and informal) are invariably held as audience input often also feeds back into performance. This has been especially true of shows directly about specific medical conditions, like "Something Somatic", about Chronic

Fatigue Syndrome/ myalgic encephalomyelitis (CFS/ME), when audience members often have direct experience of the subject matter.

Theatrescience undertakes the commissioning of scripts, productions of plays, the running of workshops and festivals and conferences. Its work has engaged schools, colleges, research institutes, community organisations, theatre practitioners, scientists and the general public. Specialist science advisors are sought for each project and client/patient groups, where appropriate are also involved (for example, CFS/ME Support Groups for "Something Somatic" and The Alzheimer's Society for "State of Nature"–see below). Much of Theatrescience's work has been funded by the Wellcome Trust; other support has come from the Arts Council, the British Council, the Peninsula College of Medicine and Dentistry (PCMD–UK), the National Centre for Biological Sciences (NCBS–India) and the Royal College of Physicians. Our partner organisation in India, other than NCBS, is the Jagriti Theatre in Bangalore. Theatrescience has worked extensively with medical students at PCMD and biology research students at NCBS. It has worked with many and various scientists and theatre practitioners in the UK and India and has connections in various other countries such as Brazil and Argentina. Theatrescience has also worked with health workers and drama practitioners in Uganda.

All our projects have been independently evaluated. Evaluation has been formative as well as summative and methodologies used have included questionnaires, personal interviews, group discussions and self-completion surveys. Such evaluation is mandatory with projects funded by The Wellcome Trust but both quantitative evaluation (numbers involved in process, actual audiences and "reach"–publications, website, etc.) and qualitative evaluation (the affective outcomes and outputs) have been undertaken on all projects to one degree or another. Many of these evaluations have been published on-line.

Our work has been represented internationally at conferences and seminars in Switzerland, India, South Africa and Argentina. Theatrescience plays have been seen in the UK, India, Uganda, Brazil and Australia. Subjects for our work have often come direct from scientific research (especially at NCBS) but have also been suggested by individual scientists ("Something Somatic") and theatre practitioners ("Bad Blood Blues"). Audience feedback has also, to some extent, changed the nature of our productions ("The Clearing", "Something Somatic", "Amol's Stories", etc.–see below).

The Work

Theatrescience's first project was the Newbury Science/Drama Festival involving six schools at the Corn Exchange Theatre, Newbury in 2002. This grew out of work undertaken by Rebecca Gould and Jeff Teare on The Wellcome Trust's "Science Centrestage" project. The Newbury Science/Drama Festival attempted to address some of the lessons learned from "Science Centrestage" in terms of process ("ownership" of product) and presentation (application of theatrical techniques). Newbury was followed by the Plymouth Science Drama Festival involving four schools at The Drum Theatre, Plymouth. This further developed the involvement of pupils in the whole process with more direct input from Theatrescience into the schools. Both these projects involved working with students, science and drama teachers and theatre professionals. The students themselves devised and performed their own plays on such subjects as Cystic Fibrosis, Huntington's disease, HIV, access to genetic medical information, etc.

Experience from these two festivals was codified into "The Drama of Science".

The Drama of Science

Experience gained from "Science Centrestage" and the Newbury and Plymouth Festivals was then condensed into "The Drama of Science". This comprised a handbook and video which were published and made available on-line. "The Drama of Science" was then rolled-out nationally to schools with workshops in Bristol, London, Southampton, Stoke-on-Trent and Manchester. Practical work was further undertaken directly by Theatrescience in schools in Winchester and Plymouth. The work was independently evaluated by Dr Penny Fiddler (2005, 12), who stated

> All teachers who created Science-Drama Projects enjoyed and learned from the experience. In general the students who took part gave extremely positive feedback about their perceptions of science. However, most of the teachers found that creating, or attempting to create cross-curricular science-drama work was extremely challenging within their schools.

Imagining the Future

This was a workshop involving scientists, writers and theatre practitioners run at the Theatre Royal Plymouth in February 2003. Its main aim was to address the perceived "conceptual gap" (in the UK at least) between theatre practitioners, research scientists and clinicians. It is our general experience that, while many British scientists might have an interest in the arts few have any practical experience, while very few theatre practitioners of our acquaintance have much interest in science at all. The situation is perhaps not as extreme as it was in 1959 when C.P. Snow (1998 [1959]) referred to it in his famous Rede lecture, but the division still exists, probably due to the resilience of our traditional educational system.

Short pieces of work were developed with actors and four of these were then commissioned as full-length plays. Two of these, "Seeing Without Light" by Simon Turley and "Still Life" by Charles Way, were later produced as part of "The Theatre of Science" (see below). Of the other two, one about eugenics was read and discussed in Sweden, while the other about cloning eventually made it onto BBC Radio.

Various long-term connections were made at this workshop with organisations such as the Peninsula School of Medicine (later PCMD) and individuals; Simon Turley (a local writer and teacher) became Theatrescience's writer-in-residence and Professor Anthony Pinching (Sub-Dean at PCMD) became Theatrescience's senior science advisor.

"Imagining the Future" became the basis for our first large-scale theatre project, "The Theatre of Science".

The Theatre of Science

This project ran at the Theatre Royal, Plymouth, as well as in various schools and venues in Devon and Cornwall, from 2004 to 2005. It included professional productions at The Drum Theatre–"Seeing Without Light" by Simon Turley, about HIV research in Africa and "Still Life" by Charles Way, about genetic predisposition–a large community event, the making of videos on genetic engineering and cochlear implants, workshops, public presentations and discussions as well as performances by seventeen schools. Local community groups were also involved including a Young Mothers Club and an HIV Support Group. The project was independently evaluated by Helen Turner of the Institute of Education and Rachel Dickinson of National Theatre, Education (2005, 18), who stated

> Everybody (education, theatre and public) was in agreement that this medium of theatre successfully engaged people with thought provoking and informative productions. Students working on their own productions found that they could learn and understand more through this experiential process. Teachers said that students learned more this way than in ordinary

Science lessons and were the strongest advocates of theatre as a learning tool.

In November 2005, Jeff and Simon presented a Theatrescience Workshop at the "Science on Stage" Conference at CERN, Switzerland. This workshop involved over one hundred educationalists from thirteen countries. It was intended that the "Theatre of Science" should then be transferred to another UK theatre base. However, instead, Rebecca and Jeff decided to transfer the model of work to India. Rebecca had already done some work in India for the British Council, while Jeff had worked extensively in British Asian theatre in London and they both thought that operating in India might sharpen and focus the company's work in a way not really possible in the UK.

Theatrescience began exploratory work in India in the summer of 2006. Meetings with writers, actors, students and directors took place in Kolkata, Mumbai and Bangalore. Workshops were also undertaken with theatre professionals in Kolkata and Bangalore as well as with students at the University of Mumbai. These formed the basis for "Imagining the Future–India".

Imagining the Future–India

This workshop/festival, in association with the Artistes Repertory Theatre (ART)/ Jagriti Theatre in Bangalore, followed on from the Plymouth workshop in 2003 and the initial work in India in 2006. It comprised an eight-day event in Bangalore consisting of workshops, readings and presentations aimed at stimulating inter-cultural and interdisciplinary scientific and dramatic collaboration. (One of the scripts developed was "Bad Blood Blues" by Paul Sirett, see below). A number of groups, schools and venues were involved in Bangalore including, for the first time, the National Centre for Biological Sciences, where a workshop was held with PhD Biology students and staff. Four of the new plays created were then brought to London as "Theatrescience India in London".

Theatrescience India in London

The Indian companies involved were The Creative Arts from Kolkata and ART/Jagriti from Bangalore. Writers involved included Gautam Raja, Farhad Sorabjee and Paul Sirett.

"The Invisible River" by Gautam Raja and "Crab Soup" by The Creative Arts were performed at the Lilian Baylis Theatre, both being followed by chaired discussions. "The Invisible River" explores the socio-political and religious effects of pollution in the Ganges, and "Crab Soup" is about women living with HIV in India. "Bad Blood Blues" by Paul Sirett about drug trials in Africa and "Idiot Wind" by Farhad Sorabjee about attitudes to autism in India received public readings at the Soho Theatre. Workshops on "Darwin and Evolution" and Verbatim Theatre techniques based on interviews with Indian women affected by HIV were also held at Soho. A presentation was made at the Nehru Centre (Indian High Commission) and "Crab Soup" was also seen at Tara Arts.

An independent evaluation undertaken by Dr Jessica Mordsley (2008, 30), stated

> In conclusion...['Theatrescience India in London'] succeeded in achieving increased public engagement with biomedical scientific issues, and made many new connections between cultures, continents, disciplines, individuals, and organisations.

Something Somatic

This new play by Simon Turley "about" CFS/ME (the condition is never actually mentioned in the script) was performed by Theatrescience in Plymouth and London in November to December 2007. It was first explored in a workshop at "The Theatre of Science" and then developed over two years by Simon, Tony and Jeff with input from a large number of CFS/ME sufferers, carers and medical practitioners. During this process it received two public readings, one at the Royal Academy of Dramatic Art, and various presentations were also made at medical conferences. It was independently evaluated by Melanie Scaffold (2008, 10), who stated

> The consistency of positive audience response across the three evaluated performances demonstrates that, while many were drawn to the play because of the issues it addressed, its success as an exciting piece of theatre is not dependent on the extent to which the audience is engaged in these themes. When questioned about what was most enjoyable over 50% of the audience referred to the 'theatricality' of the piece–set, lighting, sound, direction and acting–indicating the importance of the quality of the dramatic experience.

In June 2008 the Wellcome Trust invited Theatrescience to perform extracts from this production at a meeting of its Public Engagement advisors at the Eden Project, Cornwall (UK).

Figure 5.1. Production shot from "Something Somatic" by Simon Turley, featuring Richard Pepper and Rachel Donovan, London, 2007

In December Rebecca represented Theatrescience at the first "Wellcome International Public Engagement Conference" in Durban, South Africa. Also in 2008, Tony, Simon and Jeff contributed to the Wellcome publication *Creative Encounters* on the subject of "Dramatic Clinical Spaces". Tony (Pinching *et al.* nd, 90) stated

> As a team from varied disciplines working on the Theatrescience project, we have engaged in dialogues and discourses that have increasingly taken us from fairly straightforward issues of securing veracity towards establishing a deeper and wider authenticity. This journey should give us, and others, greater insights into the interfaces between our humanity/ identity and the processes through which we explore outer and inner worlds.

Residency at the National Centre for Biological Sciences

Theatrescience returned to India in January 2009 for a month-long residency at NCBS to develop work directly from current research being undertaken there. After a period of seminars and conversations with scientists, two areas of work were selected which were developed into scripts. One, "The Clearing" by Gautam Raja is concerned with some of the public health issues involved in South Indian forest ecology. The other, "Amol's Stories" is about post-traumatic stress disorder (PTSD), memory and Indian attitudes to mental illnesses. It was written by Jeff from improvisations by the company, directed by Rebecca. Both were performed in "scratch" versions to staff and students at NCBS and their comments were fed back into the scripts. Further versions were then presented at a venue in Bangalore itself to the general public followed by a chaired discussion. Also as part of this residency further work was undertaken in Bangalore Schools. A reading of "Bad Blood Blues" was also given at a medical ethics conference.

In March, Jeff attended the Campus Euroamerican in Buenos Aires where his Theatrescience workshop was a plenary event. It largely dealt with international perspectives on the "two cultures" (Snow 1998 [1959]) in terms of how the presence and nature of the rift between art and science differed in different countries. The presentation also included a workshop element on access to genetic medical information.

In May 2009, "Bad Blood Blues" by Paul Sirett was produced at the Theatre Royal, Stratford East in London.

Acting against Worms

In October 2009, Theatrescience began an educational science/drama project about Bilharzia, in association with Schistosomiasis Control Initiative (SCI) based at imperial College, London and the Ugandan Health Ministry in the Busia district of Uganda. Initially this was conceived purely as a public information activity, but the work finally produced was much more concerned with social and economic issues than the mere presentation of preventive information.

At a very early stage in the project it became obvious that the 400 hundred or so children involved knew perfectly well what they should do to avoid bilharzia. The real issue was why they could not or would not follow the advice they had already been given. Workshops were held with medical staff, local drama practitioners and in nine schools. Students in the schools were then encouraged to develop their own ideas which they sent to Theatrescience back in the UK. These were then worked up into scenarios which were returned to the schools. The local drama practitioners then worked in the schools to develop short plays, most with music.

The Uganda project was completed in April 2010. A Festival was held with performances from five schools to an audience of over 500 students, local people and dignitaries. The main problems with trying to avoid bilharzia which concerned the children were: the distance to the nearest borehole (safe water) and the queues when you got there; the likelihood that the borehole would be broken anyway; the necessity of fishing in Lake Victoria for food; and the price of gumboots.

Also in 2009, a reading of "Bad Blood Blues" was given at a medical ethics conference in Bangalore, while Rebecca and Dr Jessica Mordsley (Theatrescience Associate) curated the "Second Wellcome International Public Engagement Conference" also in Bangalore. This brought together many educationalists, scientists, artists and NGO workers at NCBS to engage in a variety of themed activities based around "Narrative" in its various forms and meanings. Theatrescience (Jeff Teare, Simon Turley, Paul Sirett and Arundhati Raja from Jagriti) delivered two workshops, one about process, the other comprising of extracts from previous Theatrescience shows and a general discussion of "sci-art".

In 2009, Jeff received a "Darwin Now" award from the British Council to develop a piece about the influence (or not) of Darwin on Indian thought and culture. This involved two research trips talking to a wide range of Indian scientists and religionists. In June 2010, the resultant play "Darwin in India" received a rehearsed reading at NCBS followed by a discussion with research scientists. This event has proved crucial in the development of Theatrescience's thinking on the social-cultural aspects of the company's work in India and fed directly into "The Ashtavaidya Play" project (see below).

In June 2010, "The Clearing" by Gautam Raja, first developed at the NCBS Residency, was produced at the Alliance Francaise, Bangalore by Jagriti.

Not My Fault

This major programme of work grew out of the presentation of scenes from "Something Somatic" at the Eden Project in Cornwall in 2008. It eventually involved six writers, ten scientists and ethicists, fifteen theatre practitioners, over thirty students, at least twenty Eden Project staff and hundreds of the general public. It was funded by The Wellcome Trust under their "Genetics and Health" programme.

The project started in November 2009 with an initial workshop involving scientists, clinicians, ethicists, writers and theatre practitioners. This was held in the Eden Project. Seminars were held, previous Theatrescience work was discussed and an extensive tour of Eden was undertaken as the final performances were to be site specific and influenced. Six scripts were then commissioned on various aspects of the subject "genetic variance and health". Three of these commissions were with established playwrights and three were with young writers as part of a development programme.

In July 2010, as the first performance event in the "Not My Fault" programme, students from Plymouth's Ridgeway School presented their devised piece about genetic inheritance and health at Eden. Then, in November 2010, despite severe problems with flooding, "Not My Fault" culminated at Eden with the three short plays by young writers being performed in the Temperate Biome and Simon Turley's "State of Nature" (about Lewy Bodies Dementia) being presented in the Tropical Biome (this was also performed at The Lost Gardens of Heligan and Truro Hospital, with a try-out at The Pleasance Theatre in London). Part of the idea of performing at Eden was to reach a potentially "non-theatre going audience". The three short plays were presented as "guerrilla theatre" with little or no pre-announcement and even "State of Nature" was seen by many Eden visitors who had no idea they would see a piece of theatre when they first arrived.

Medicine as Art

From 2006-2011, Jeff taught at the Peninsula College of Medicine and Dentistry, under the Theatrescience banner, a short "Medicine as Art" course about how doctors' attitudes to disease and injury might affect their clinical practice. The project looked at how doctors' religious, moral or familial concerns might impinge on their treatment of patients given such conditions as unwanted pregnancy, early onset diabetes, cancer and CFS/ME. Students were encouraged to express their own concerns, which were especially relevant to one or two who suffered from chronic conditions themselves.

This work was suggested by Tony Pinching and resulted in over 40 short pieces of drama (mainly live but some video) reflecting the students' attitudes to various medical conditions. A stand-out event was the performance of a short play about irritable bowel syndrome in a gents toilet. These pieces were written and performed by the medical students involved, apart from one year when professional actors were engaged. They were presented to and discussed by other students and staff of the College.

The Ashtavaidya Play

In September 2011, Jeff became a Visiting Scholar at NCBS to create a piece of drama about the place of traditional medicine in modern India. Specifically, the project looked at research undertaken at NCBS into the current state of Ashtavaidya medicine (the Keralan form of Ayurveda) and how it might, or might not, be fully incorporated into modern Indian medical practice. Current research at NCBS seeming to show efficacy in some Ayurvedic treatments were considered, as were the social and cultural role of such treatments as opposed to the purely medical. The drama was devised and performed by PhD Biology students. Three performances were finally given, one to other students and staff, one to the general public followed by a panel discussion, and one at an Ayurveda Research Centre.

Additionally in 2011, "Bad Blood Blues" by Paul Sirett was performed to critical acclaim in Rio de Janeiro and Sydney, while in 2012, activities by Theatrescience included: mounting a shortened version of "Something Somatic" by Simon Turley at the National BACME Conference in Milton Keynes (UK); running a three-day Theatrescience Workshop with Oxford School of Drama students in London; introducing Theatrescience techniques to students on the World Theatre Course, East 15 Acting School at Essex University; delivering a research seminar on Theatrescience at Exeter University; and reviving "The Invisible River" at the new Jagriti Theatre in Bangalore.

Development

Over ten years of work "bringing science and theatre closer together", as stated in Theatrescience's first "mission statement", has resulted in a large number of performances of a wide range of work from a "Science Cabaret" in Plymouth, through fringe performances in London, an open air festival in Uganda, various shows in Indian venues from research institutes to brand-new theatres, to major theatre presentations in Rio de Janeiro and Sydney. Hundreds of theatre practitioners, scientists and medical practitioners have been involved, thousands of students and tens of thousands of audience members. All of this has resulted, as it should, in major changes to how Theatrescience works. Each project has fed into a developing practice and evaluations have contributed to an internal, and sometimes external, debate into how we do things–and indeed why.

However, it has to be said that this process has not always been entirely conscious or planned. For example, one of our regular workshops on genetics has probably been delivered fifty or so times with participants ranging from teenage school children in Plymouth and Bangalore to international scientists and educationalists in Geneva and Buenos Aires. Version forty, for example, was very different to version one but as these workshops are very seldom recorded, or even noted in detail, how the workshop has evolved is often more a matter of chance and accident than planning.

Theatrescience began in 2002 with projects in schools largely as a response to working on the Wellcome Trust's "Science Centrestage". We felt it important that the students involved in such projects, often not of their own volition, should at least "own" the work they presented themselves rather than perform something imposed, however creatively, by staff or outside drama specialists. The shows had to be about science issues that actually concerned the students. For example, in one school a project initially about HIV in an African context turned into a show about a sexually-transmitted disease (not HIV) that the pupils had immediate experience of and about which they wished to educate their parents. Another school did a piece about cystic fibrosis as there were two students at the school who suffered from the condition. A few others looked at various aspects of the somewhat specious subject of "designer babies" with special reference to potential sexual attractiveness and sporting prowess.

We also insisted that every school involved had both a science and a drama teacher on the project. On more than one occasion, I heard a science teacher say that they had never been in the school's drama studio before and drama teachers admit that they had never even thought of taking a science subject as the basis for a devised drama piece. The best work produced not only did bring science and drama closer together, often very physically, but also actively engaged students with current societal, political, ethical and often personal moral issues.

We have also taught a bit of science along the way but this has never been our main intention. The participants need to understand enough science in order for the drama to be accurate, but beyond that our concern has always been how the science affects society and the individual. However, the potential kinaesthetic learning and general educational effects of our work should not be downplayed too much. Anecdotally, one student was once heard to say after a workshop, "I learnt more about biology today than in the last three years" and a science teacher observed on video that participating in a drama project helped him understand his pupils better.

Working in schools was too narrow a focus for us as theatre practitioners, so with "Theatre of Science" we began working with adult groups and commissioning playwrights to write plays for general audiences. These plays were then published and given a further life through the website and presentations at conferences, etc. We then came to think that performing mainly in theatres, or to special interest groups at conferences, was also too limiting. This led to performing to visitors at the Eden Project, many of whom were not expecting to see a play at all, let alone one on a subject like dementia.

The move to India, as mentioned above, was to do with testing our process in a different, "sharper" cultural context. Some of the work produced, especially the Jagriti productions of Gautam Raja's scripts, has been artistically outstanding, but our other work there especially at NCBS made us realise that we had to learn some lessons about cultural imperialism in general and different attitudes to the performing arts in particular. At a very basic level, we met more scientists in India with an active artistic practice in drama, music or dance than we ever have in the UK–and a "Brit" cannot work in India without being forcibly reminded of certain aspects of the Raj which they would prefer to ignore.

In one way, the last project in India, "The Ashtavaidya Play" at NCBS, brought our work full circle in that the work was devised and performed by the students themselves, this time by PhD Biology students. The "danced poem" in the performance (viewable on our website video) suggests that the divide between art and science may not be as distinct in India as many think it still is in the west. These cultural resonances of our work are definitely something we wish to pursue if and when we succeed in raising the funds for another project in India.

The funding situation, both for art and science, becomes ever more difficult in the UK. Nevertheless, at the time of writing, Theatrescience is embarking on a major new project in association with the Royal College of Physicians looking at the defining moments in the early stages of doctors' careers, when they first experience a sense of ultimate responsibility for another person's life. The final "dramatic construct" created from research, the form of which is not yet known (our process is somewhat experimental after all) is scheduled to be performed in late 2013 or early 2014.

For more current information please visit: www.theatrescience.org.uk.

References

Fiddler, Penny. 2005. *The opportunities and challenges of "embedding" biomedical science-drama projects in secondary schools. An evaluation report for The Wellcome Trust-funded "Drama of Science" Project*. http://www.theatrescience.org.uk/ee/index.php/media.

Mordsley, Jessica. 2008. *"Imagining the Future III: Theatrescience India in London" external evaluation*. http://www.theatrescience.org.uk/ee/files/reports/ITFIII_Evaluation.pdf.

Pinching, Anthony, Jeff Teare and Simon Turley. nd. Dramatic clinical spaces. In *Creative encounters: new conversations in science, education and the arts*, 90-110. London: Wellcome Trust. http://www.wellcome.ac.uk/About-us/Publications/Books/WTD0 28351.htm.

Scaffold, Melanie. 2008. *"Something Somatic" external evaluation*. http://www.theatrescience.org.uk/ee/index.php/media.

Snow, C.P. 1998 [1959]. *The two cultures*. Cambridge: Cambridge University Press.

Turner, Helen, and Rachel Dickinson. 2005. *"Theatre of Science" external evaluation*. http://www.theatrescience.org.uk/ee/index.php/media.

CHAPTER SIX

LOOKING BEYOND NEEDS: CAPACITY FOCUSED DEVELOPMENT THROUGH THE SOCIO-DRAMA TOPOGRAPHY PROCESS

STEPHEN SILLETT AND JENNIFER JIMENEZ

Introduction

Typically, Community-Based Organisations (CBOs) are most connected to their communities through the services they provide. The participatory development approach shared in this chapter builds strategic capacity in a CBO to learn from and respond to the direction given by the community members they service. This capacity focused development approach integrates arts-based methodologies that facilitate relationships between communities and the local CBO. Through such arts processes, deep insights grounded in the lived experiences of community members are created, shared, honoured and developed.

CBOs often deal with complex social change where unique local realities arise from an interaction of multiple influences, creating an unpredictable, uncertain environment. Our approach seeks to take a holistic view to development with community as navigator of this shifting multidimensional reality. We will speak here to the capacities required to implement projects using a process designed to facilitate active inquiry and location-based conversations with local communities around their priority issues.

Reflective and dialogical art processes play an important role in community-engaged projects. This is related to increasingly interactive practices in Visual Arts, Community Arts and Theatre for Development. These creative forms of expression allow people to share experiences, which can support community building and inclusion among community members. Aiding Dramatic Change in Development (ADCID) has integrated these approaches into development and community empowerment programmes leading to the creation of a unique reflective and dialogical arts-based inquiry process called Socio-Drama Topography (SDT). The SDT process engages large groups of 20 or more participants. It is designed to reduce barriers to participation faced by marginalised communities including those relating to varying levels of literacy. Through this multi-day process these groups collaborate in deep, open and strategically relevant conversations.

During this chapter we present examples of engaging groups through the SDT process in two applications:

1. Formative period for SDT. Peer Influence SDT covers a critical pedagogy and participatory theatre intervention that ran for over 5 years in schools. It is from this behaviour change project addressing HIV/AIDS and teen pregnancy that the SDT process emerged as a multi-day, arts-based inquiry process.

2. Case example of SDT used in community development. WASH SDT involves adapting the SDT process for engaging adult community members around their local Water Access Sanitation & Hygiene (WASH) issues. This project has reached the pilot stage of engagement, which has been running for over a year. We will speak to the implementation strategies for the pilot, the capacities developed and the discourse in the community enabled by the interventions.

This chapter shares our practice of integrating the SDT process into long-term Community Development Initiatives and moving CBOs beyond their current engagement capacities. The discussed theory and practice around SDT concerns itself primarily with SDT for capacity-focused community development. SDT is also applicable in contexts such as behaviour change in schools, group healing, and working environments. These other applications have particular ethical and capacity considerations that are outside the scope of this chapter.

Enabling Community to Engage in Local Development Discourse

The development sector has been questioning how organisations can respond to shifting the focus of interventions away from "products" that can be easily delivered to a practice that responds to what the situation demands (Kaplan 2000). In 2007, the IKM Emergent Programme (Beardon 2012) brought together leading practitioners over 5 years to identify challenges and opportunities for change in the development sector. They heard much talk about complexity, rights and social change, but found people planning as if change was linear and predictable. This raised questions as to why the development sector plans and acts as if the world were simple, when it is well known to be complex. Their exploration around "awareness of complexity and emergence" looked into contradictions in development practice and the pressures to conform to, and even replicate, structures which sustain false certainty. They looked at innovative development strategies that could shift practice towards programmes being implemented around an awareness of complexity and uncertainty. They found that programmes that achieved some of the capacities for realising a vision of adaptable development focused on supporting local participatory processes. They also noted the negative effects associated when "expert" or "technical" knowledge is introduced to marginalised communities, dominating local development discourse and undermining local knowledge.

This chapter seeks to share processes that enable a CBO to better understand how community assets can be used to address social and economic challenges. By capacitating community-based inquiry on an issue of concern, a mechanism for bottom-up organisational learning with the community as valuable resource can develop. This does not mean downloading responsibilities for service provision onto the community that should be provided by municipalities or the CBO, but rather, shifting the role of community members, including those with substantive needs, to a position where they are also seen as valuable resources, whose hopes and efficacy can make change happen.

ADCID believe there needs to be a change in development practice, where local CBOs take a more empowering role, with community members moving beyond clients of services to producers of change. For this empowerment to happen in a meaningful way, the community has to be perceived, engaged, and capacitated in a radically different way. By focusing on enabling capacities for local community members to explore issues and engage in strategic conversation, we propose that CBOs can form holistic strategies that empower, nurture and sustain local innovation. Staff at a local CBO can then move beyond having the capacity to administer services and deliver messages to develop more broadly what Robert Chambers (2012) refers to as "the power to empower". We believe it is through an integration of services provision, in conjunction with the facilitation of group inquiry and empowering relationships that more sustainable conditions for change emerge.

Working with complex social issues requires the capacity to have an awareness of the whole environment in which you are situated. Navigating interconnected community issues in a holistic way necessitates a multi-level systems thinking perspective. At its simplest level, systems thinking is the process of understanding how things, regarded as systems, influence one another within a whole. It uses a framework that is based on the belief that the component parts of a system can best be understood in the context of relationships with each other and with other systems, rather than in isolation. Organisations are often looked at as being composed of staff following internal procedures that connect together to form a closed system. However, when looking at organisations adapting to the external environment and responding to the changing developments in the community, then open systems thinking is more appropriate. To allow organisations to engage system wide issues of importance, a range of group methods have developed. These methods are utilised in Large Group Interventions (LGI) to provide organisations with a collaborative approach for putting open systems thinking into practice as strategies to enable change.

LGIs are collaborative inquiries into organisational systems, practices and processes that are designed to create alignment around strategic direction and system-wide issues (Purser and Griffin 2008). Current literature shows LGIs as having found widespread traction in organisations (Bunker and Alban 2006) and more recently in community contexts such as the Transition Town movement and the New Economics Foundation. These multi-stakeholder processes aim to "get the system in the room", to address change and draw on open system thinking approaches. Unfortunately, multi-day interventions are rarely used where marginalised communities are positioned as the agents of change. Even the integration of LGIs into pluralistic social technologies such as Design Labs and Change Labs have concentrated on design experts and local leaders. Where community members are featured, it is primarily as ethnographic sources for the labs to draw understanding. However, there is recent active research to find ways in which Social Innovation Labs can be developed with direct involvement of all sections of society in the process.

The SDT process seeks a location-based systems perspective and brings the system, referred to in SDT as the social landscape, into the workshop. The system in SDT is experienced physically through a unique mapping process integrated with visual and socio-drama processes that explore the social landscape relating to topics under consideration. It is through the mixture of visual and embodied processes used to feel the social landscape that a deeper understanding of the contextual factors in challenging environments is possible. Central to ADCID's work and the SDT process are dramatic approaches that as Nicholson (2005, 129) states

> enable people to move beyond the ordinary and everyday and use the aesthetics of drama, theatre and performance to gain new insights into the social and cultural practices of life.

The Socio-Drama Topography Process

SDT engages a group of 20 or more participants in a multi-day exploration using immersive arts-based inquiry. The SDT is an experiential journey where the participants are orientated to the theme through non-verbal dialogue and collaborate to create a collective map of the physical areas that relate to the topic of exploration. Using artistic materials and thematic dimensions, participants freely relate to the map and make explicit areas in the landscape, where they sense personal or collective meaning. Once complete, participants explore the physical, social and emotional landscape of the map, being open to where they feel an embodied resonance to what is presented and spontaneously group at locations on the map. These groups then follow a process to co-create physical theatre and socio-drama using the map as inspiration. While the map represents the area being explored, it is through artistic expression of the social landscape and thematic annotations that a more diffuse understanding of the social topography overlaying the map emerges.

The interactive first three stages of the SDT workshop reveal narratives and theme related information, which can be reflected upon and analysed in the fourth stage. The empowering strength of the fourth stage of the SDT process is that it readies participants for deliberative analysis involving reflection anchored by the thematic and social landscape (the map) explored in the first three stages. Participants reflect on the meaning co-created during their earlier exploration and propose possible actions to address the issues raised. These reflections generate ethnographic information that can inform the organisation and other development actors. Participants have the opportunity to form broad system wide connections through sharing their collective experiences and reflecting on multiple perceptions around an issue. Because this situated inquiry has developed a depth of reflection, SDT is able to act as a proposal forming activity to enable community development and innovation. These process outcomes are unique to each project and depend on the immediate and long-term aims, issue of exploration and expectations of participants.

By collectively sharing experiences through embodied exercises and performed stories, and deconstructing and discussing the social implications and circumstances around these images and stories, the group co-creates knowledge. The dialogue during the workshop follows a Bohm dialogue approach, focusing on deep listening and insights that result. Bohm's thinking about dialogue was based on the belief that if people became more aware of their collective thinking and the subsequent meaning it produced, it would coherently move them towards a sense of the whole and allow them to transform their position for the better (Ellinor and Gerard 1998). For Community Development applications of SDT, Bohm Dialogue is used throughout the first three stages of the SDT experiential workshop process, leaving the fourth stage of the process to follow with more deliberative approaches. In school-based applications of SDT other dialogue and critical analysis methods are used.

While the discussion is deliberative in the fourth stage, it is seeking to find proposals for moving forward, not consensus. This is to ensure that the exploration and creative proposal forming intent of the workshop is maintained. It is by going through this orientation process in a free and open way that the intangible capacities are developed in the group members. The larger project can have other activities that follow on from the SDT process and connect to other groups and development activities supported by the CBO and their partners.

Adapting SDT to Explore Different Themes

The SDT process is suited to local issues that are complex and/or uncertain and in which a group of local community members have a shared relationship. A theme is chosen to enable the group to explore the issue and find concrete local meaning. Each priority issue around which change is desired requires a thematically adapted design of SDT. The pluralistic approach to adapting the SDT process to a new theme and project purpose is organised by ADCID practitioners, in collaboration with local community members and CBO staff. Understandings around the process design occur during the training and trialing period. Once the process design is ready for piloting in a live community programme, a rolling Developmental Evaluation is initiated. A Developmental Evaluation informs and supports innovative and adaptive development in complex dynamic environments. Developmental Evaluation brings to innovation and adaptation the processes of asking evaluative questions, applying evaluation logic, and gathering and reporting evaluative data to support project and organisational development with timely feedback (Patton 2011). This form of evaluation is suited to both the larger project development and SDT process adaptation.

Working with a Community Hub

When engaging with marginalised communities, the broader container and strategy formulation of an initiative is a key responsibility of those organising the intervention. There are considerations that need to be taken into account before embarking on using a deeper engagement process such as SDT. These ensure that as well as being transformative the SDT process is contributing to a larger change initiative.

When starting work on the ground with a partner organisation, we recommend spending a minimum of two months immersed in the community context and observing the organisational operations to sense local conditions and feasibility for engaging the community through an SDT-based approach. In South Africa, ADCID has spent over two years developing the partnerships before trialing an SDT Initiative. This builds understanding of the capacities of a local partner organisation and their existing relationships with the community. When a suitable opportunity arises, the next stage is to develop with local front-line staff and community contacts details of complex social issues, which could benefit from an SDT intervention. Key qualities of our partner CBOs include:

- Leadership with a vision for community to develop their social and practical abilities
- Ability to assimilate ways of working which respond to community capacities for development
- Existing relationships in the community
- Sufficient infrastructure to handle core services
- Culture which values critical thinking and listening to community members
- Interest in sharing knowledge and empowering other local organisations
- Ethical framework for staff engagement with community members.

SDT Process for Community Development toward Innovation

The community-based partner with an interest in asset-based development becomes the local hub in the initiative. The local hubs take ownership of the initiative providing operational resources and staff.

ADCID collaborates with the hub organisation on programme strategy and building capacity to enable implementation (Figure 6.1).

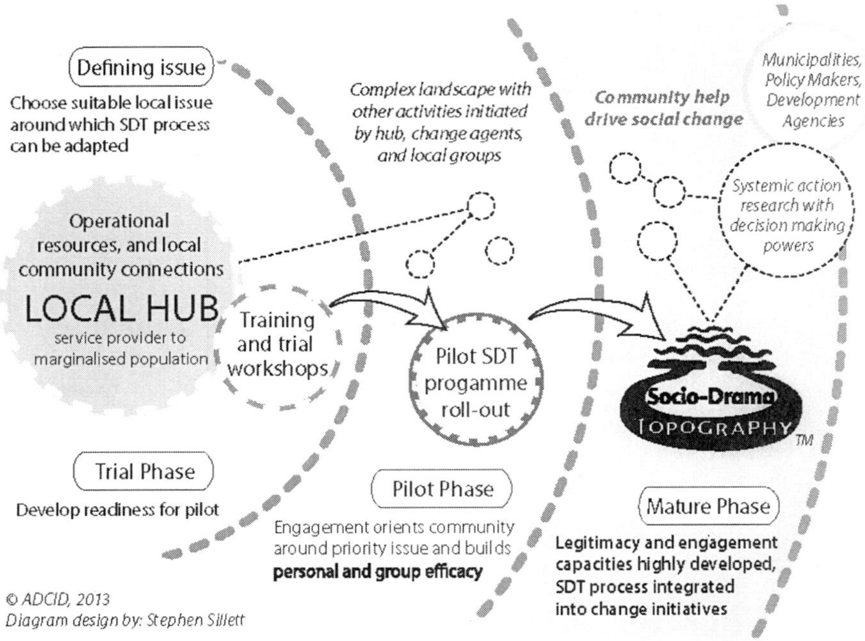

Figure 6.1. Phases of an initiative using Socio-Drama Topography (SDT) process

In the first (Trial) phase, local volunteers and staff experience drama workshops and training in participatory exercises and the basics of the SDT process. Simultaneously, the staff and volunteers share insight into the local context and theme exploration. This informs the workshop design and the initial adaptation of SDT on the specific theme.

In the second (Pilot) phase, a pilot engagement is planned and rolled-out with community members participating in multi-day SDT workshops. The Pilot workshops build relations in the community by bringing members together to engage around a relevant theme. These workshops are part of a complex development landscape and run with other activities initiated by the hub, change agents and local groups.

In the third (Mature) phase, SDT workshops are adapted into a mature form and integrated into initiatives based on community driven solutions and innovation. During this phase the developed programme/initiative has gained legitimacy in the community and networks between municipalities, policy makers and development agencies are developed. Deliberations in stage four of the SDT process become a strategic vehicle to connect participant proposals with policy and decision-making powers helping the community to drive systemic change. If the social and organisational landscapes of the NGO are seen in terms of systems, then with a capacity focused project, ADCID aims to intersect these and to move the system to different states using SDT and other processes as a catalyst for change. Over a longer period, the aim is for the system to reach a state whereby community members have the confidence to innovate and contribute their abilities and insights, and the organisation has sufficient ability to respond with emerging strategies to develop wider systemic action and practical initiatives.

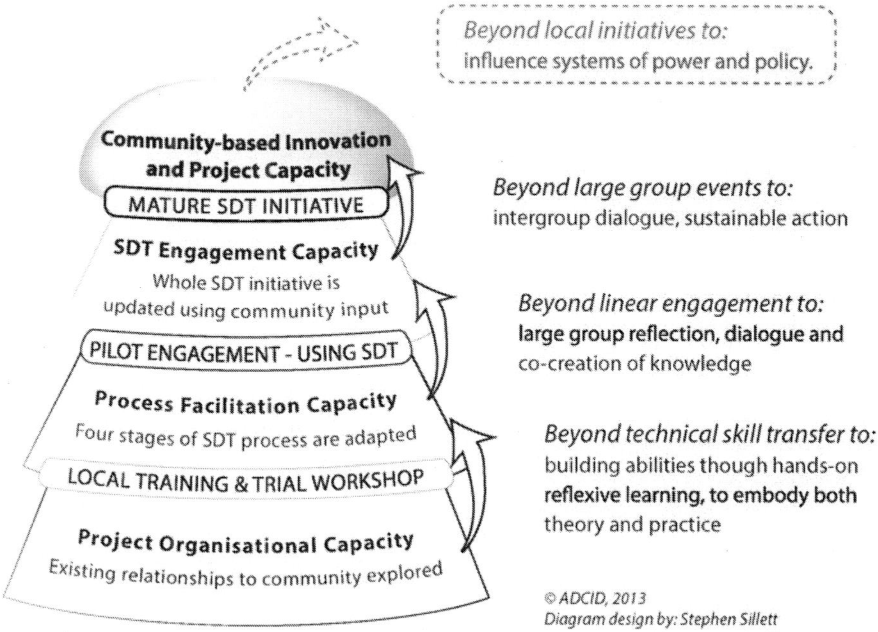

Figure 6.2. Stages of readiness for a project utilising the SDT process and capacities achieved through external technical support

Figure 6.2 shows the stages of readiness for an SDT initiative, working toward the long-term aim of community based innovation at the top. The approach is capacity focused and progresses as increased infra-structure, facilitation and ethical requirements are attained. Assessment regarding whether a state of readiness has been sufficiently achieved to move to the next phase should be decided in relation to each particular project, local situation and theme.

Building Process Capacities

Initial training involves the general use of participatory, experiential and drama based techniques and philosophies that can be useful to the organisation. These activities lay the groundwork for discussing how a capacity focused SDT intervention maybe adopted by the organisation. A key aim at this stage is to inspire staff around the potential of experiential approaches and build motivation around engaging the community in a more participatory way. For some this may involve unlearning normalised approaches and also addressing powerful internal resistance to what may be seen as less modern and scientific than the technical development tools commonly used, such as logic frameworks and needs audits. By experiencing the exercises themselves through these training sessions, staff and volunteers experience the insights that can be evoked by community members during engagements and also the vulnerabilities and confusions that are common when exploring in physically embodied ways.

Once the CBO and ADCID commit to addressing a priority issue using an SDT intervention, ADCID forms an architecture for training local facilitators in the SDT process. This includes hands-on practice of facilitating groups using interactive drama approaches. The training is structured with the aim of enabling local facilitators to run SDT workshops at levels of engagement and complexity that are appropriate. This is when training moves beyond technical skill transfer to building abilities through field learning and embodying attitudes and behaviours that seek to empower.

Process skills and development of local arts-based facilitators evolve over time. The scope of the project is set at an appropriate level based on the growing capacities of the facilitators and the engagement capacities that are “opened up” during the first pilot workshops. A primary focus of training is the understanding of differences between didactic teaching, skills training and the use of open inquiry and reflective facilitation processes.

From Pilot to Broader Community Engagement Using Mature SDT

During the SDT Pilot Phase, capacities are established to enable a broad community engagement initiative. Levels of trust between the hub organisation and community are deepened. The community builds an appreciation for participatory approaches, which generates interest in concurrent and future initiatives. Community-leaders are identified who can be engaged with further activities. Increased understanding around the theme of inquiry fosters interest into further exploration and follow-up activities. The community is valued and welcomed to participate in innovation and suggest designs for change.

CBOs interested in going beyond the pilot phase of an SDT intervention can integrate the SDT process into a broader engagement initiative. This would take the CBO beyond running large group events that open up engagement and closely integrate the SDT process into a wider development approach. In doing so, the potential for intra-group dialogue and sustainable local action is given more structure and purpose.

Emergent learning on the theme and through the implementation of the pilot phase informs this broader initiative. The SDT workshop is adapted into a “mature” version, which aims to strategically connect the experiential workshop and deliberative stages with particular change objectives happening with other development actors in the community. Existing networks with the CBO and within the community are strengthened, and new networks are established for connecting to policy and decision makers. When entering the mature phase, ADCID provides additional training which results in accreditation of facilitators in relation to that particular adapted SDT process.

The following case examples discuss two community initiatives using the SDT process. Both communities face similar socio-economic challenges and are located in rural KwaZulu-Natal, South Africa. The first example discusses a youth behaviour change project in schools and the second relates directly to community development. The SDT process is at different stages of development in each example. In both cases, ADCID worked with the local organisations to envision a way of addressing a priority issue in the community. The organisations committed resources and ADCID assisted with the strategic framework and training, enabling conditions to engage community members. The case examples outline the organisational and community capacities developed, and the emergent learning that informed and shaped the initiative.

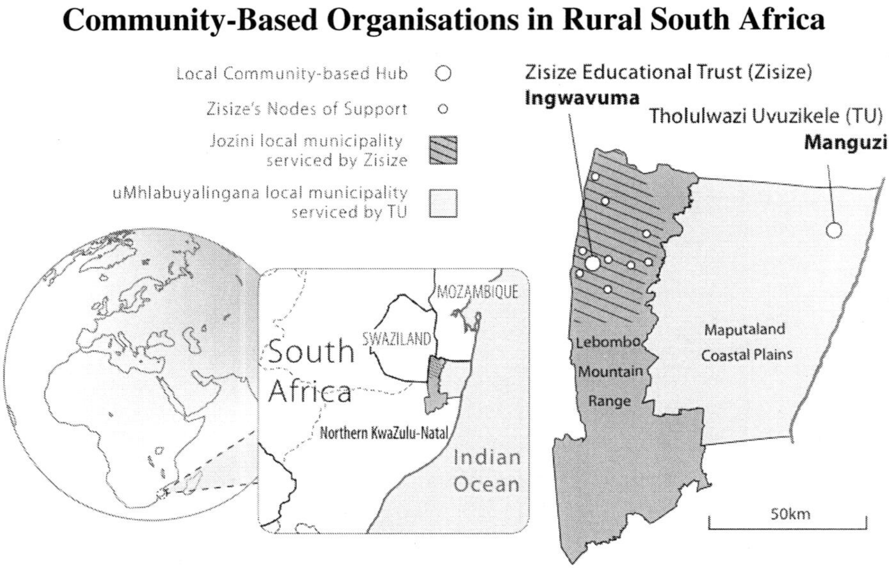

Figure 6.3. Location of partner organisations in Northern KwaZulu-Natal, RSA

The two CBOs are based in uMkhanyakude, the northern most district of KwaZulu-Natal, South Africa, servicing isolated communities. Zisize Educational Trust (Zisize) is based in the Lebombo mountain range and services "nodes of support" around the Ingwavuma region. Tholulwazi Uzivikele (TU) offices are located in Manguzi and services the local municipality located on the sandy Maputaland plains that stretch up the coast to Maputo in Mozambique. The region of KwaZulu-Natal has the highest HIV prevalence rate in the Nation with 39.5% of 15-49 year olds living with HIV (UNGASS 2012). Unemployment is extremely high in the region with unemployment rates of over 60%. This makes most households dependent on survival through subsistence agriculture and social security grants. In the uMkhanyakude District, a 2007 survey showed that only 18% had completed high school (Stats SA 2008). Oxfam (2011) states there are approximately 64% of people in need of water and 96% in need of sanitation. The district also has fluctuating high levels of malaria and cholera principally due to poor water supply, poor hygiene, inadequate sanitation and the pollution of water bodies with faecal matter. Pongolaport dam is the primary source of water within uMkhanyakude, however this is used primarily for agriculture despite statements at a

KwaZulu-Natal Water Summit assuring the population that reticulation plans would be prioritised.

Formative Period for SDT: Peer Influence among Youth in the Context of a HIV/AIDS Pandemic

The Zisize Drama Project ran from 2006 to 2012, reaching over 2500 learners and 600 parent/carers, engaging them in critical reflection on healthy choices and relationships. The engagement helped Zisize build relationships and strengthen its presence in the school communities. These relationships helped facilitate the introduction of school mentors and peer education programmes in local schools.

Zisize Educational Trust (Zisize) started in 1998 as a Teachers' Resource Centre to address the poor quality and lack of resources in the local education system, one of the apartheid legacies. In 2001, Zisize broadened its mission to include a children centre and was actively working to reduce poverty and mitigate the impact of HIV/AIDS in the Ingwavuma region. They pioneered a holistic method of service delivery using schools as "nodes of support", which has been utilised elsewhere in South Africa. They support orphaned and vulnerable children, and address child poverty through school feeding programmes, garden projects, and paying school fees. Through their two children's centres, they operate a library with tutoring, youth sports and activities, a radio project and crèche to support women and young mothers. In addition to meeting these basic needs in the community, Zisize also runs extensive Life-skills projects that range from focused psycho-social interventions to directly targeting youth vulnerabilities to HIV/AIDS.

In 2003/2004 Zisize was making provisions for increasing numbers of orphaned and vulnerable children, a direct impact of the HIV/AIDS pandemic in the region. Prevalence rates had risen, as had teenage pregnancies and levels of fear and social stigma around HIV/AIDS in the community. During that period, we volunteered for Zisize, running drama classes at a local primary school and after school sessions for teenagers over a five-week period. Conversations with the teens revealed a great deal about the pressures being felt in the rural areas. It was clear that personal identity and what it meant to be a real Zulu man weighted heavily on the male youth. Boys were strongly influenced by the social pressure to have a child by 18 years old, even though they did not feel it was the right age. Influences for girls around sexual relations ranged from wanting to be loved, to being accepted by their peer group, or getting money to buy clothes. The group expressed a deep sense of being overwhelmed by the cultural forces in the area and the fear of AIDS in the community. They wanted to move to action following the drama sessions and expressed a desire to develop the work into a project where they could engage their peers using drama as a mechanism for change.

Zisize was quick to support the youth in this endeavour as teen pregnancy and the dilemma of "children having children" was becoming increasingly prevalent. Besides the socio-economic impact teen pregnancy was having on children's lives, it indicated that there was an alarming amount of youth engaging in unsafe sexual practices in an area with the highest HIV/AIDS prevalence in the nation. Over the following 2 years, ADCID worked with the youth, Zisize, and other educators to devise a project to engage youth around issues of peer influence and healthy choices.

The Containing Project: A Radical Pedagogy and Interactive Forum Theatre Intervention in Schools

Engaging youth in discussion on healthy choices through forum theatre was an integral part of The Zisize Drama Project. A number of participatory theatre projects have been developed in South Africa in response to the HIV/AIDS struggle. These projects, many of which are performed in school environments, involve community participation through a problem-posing methodology. DramAidE (Drama in AIDS Education) based out of the University of Zululand, South Africa, uses forum theatre to encourage participation by creating dialogue and stimulating critical consciousness. Through these theatre interventions, community members are drawn into discussion regarding issues around HIV/AIDS and are encouraged to find solutions that are relevant to their lived experiences and the contextual factors that influence their behaviour choices (Govender 2011). These programmes tour local schools near the urban centres, but are less likely to visit schools in remote rural areas.

Forum theatre is one of the participatory theatre methods used to engage marginalised people developed by Augusto Boal and heavily influenced by Paulo Friere's Pedagogy of the Oppressed in the late 1960s (Nicholson 2005). In forum theatre, a dilemma is posed to the group in the form of a theatrical scene, which usually has a negative outcome. Participants are asked to step into the play and take on the role of the main character to try to change the outcome. They are invited to imagine new possibilities and solutions, and to actively try to enact the idea in the moment. Active dialogue and at times debate ensues between participants and audience members, that is supported by a facilitator referred to as a "wild card" or "joker". This interactive session is often referred to as providing a rehearsal for life.

Strategic decisions were made around the design of the engagement in order to foster adequate levels of trust and reflection from the learners to enable active involvement in the forum theatre performance and discussion. During previous visits volunteering in the community, we observed that students were accustomed to a more didactic learning experience and typically displayed low levels of confidence and efficacy. The legacy of apartheid had reduced opportunities for critical thinking, which was evident in the way the youth responded to inquiry. When asked questions, they were keen to tell you what they thought you wanted to hear rather than what they might actually have been thinking. Deeper probing and reassuring would often bring a different answer than the first "right" answer. Youth had also begun to learn the catch phrases which were part of the National HIV/AIDS campaigns. They were quick to use the terms "A.B.C." ("Abstain, Be faithful, Condomise"), yet they did not reflect a critical understanding of how these mantras applied to their own behavioural and social context. It was decided that a three-day participatory theatre and reflective dialogue workshop to introduce and deepen exploration into the theme, and build the confidence of the learners would precede the forum theatre performance. It was from the long-term development of this three-day workshop process that the SDT process emerged.

Creating a Space Where Youth Felt Safe to Explore Without Fear of Judgement

The three-day workshop was designed as an expressive and reflective inquiry process for participants to explore and discover. The learners were positioned as the expert in their lives. Great importance was placed on the power dynamics between those running the workshop and the youth participants. If the interactions between the participants and the facilitators were inscribed with a sense of "us and them", then the chance for exploration and student collaboration would become seriously impeded. Learners should be free from feelings of judgement or fear of repercussions that are inherent in teacher-student dynamics. It was decided that the facilitators for this project would be young adults from the local Ingwavuma area, who had grown up in similar social, cultural and historical contexts to the youth participants at the schools.

The project facilitation and drama capacities were developed through free drama lessons offered over a period of three months; lunch and transport was provided and sessions took place five days a week. Fifteen local unemployed adults and youth attending these sessions helped devise the forum theatre play and provided feedback on the various participatory drama exercises. The play and three-day workshop content was a reflection of the co-creation of knowledge generated among participants. At the end of this period, there were six local young adults and three youth who were consistently attending the sessions and rehearsing the forum theatre play. The three male youth were part of the original youth group in 2004 who inspired the project. The adults received an additional month of training focused on facilitating reflective dialogue and practicing sequenced activities in the three-day workshop. Time was spent to distinguish between the common-practice of "preaching" what people should do, to "listening" in a reflective way to encourage critical inquiry. After this training, six facilitators began trialing the workshop in schools, where they would be joined by the three youth who performed in the forum theatre play.

ADCID's work places great emphasis on the space in which engagement occurs and from this environment weaving in creative meaning-making processes. When visiting schools to run the workshops and performance, the traditional classroom space was transformed into an open reflective space by changing the power structures in the room through the facilitators, use of language and through the arrangement of the physical space. Conscious of the way language influences power, it was important that the facilitation and discussions happen in the participants first language, isiZulu. Learner-facilitator ratios during the workshop were significantly lower than teacher-learner ratios in the classroom: 3 to 30, compared to 1 to 60. Desks were cleared out to eliminate a sense of the front of the class and one-way flow of knowledge. During the forum theatre, the learners surrounded the action in the play (see Figure 6.4).

The first year pilot involved the local team of facilitators and actors touring seven schools. Based on learning around this first year of implementation and the community response, adaptations were made to the project. During the second year of the pilot, facilitators received additional training, with particular attention being given to competencies around spatial awareness during facilitation and their acting performances in two newly created plays. Some parents and carers had expressed a need for more information and input over the nature of the workshop discussions, as many issues relating to sexual relationships are rarely discussed openly in rural areas of South Africa. This led to the addition of a Parent/Carer event that would take place the day before each school workshop. During this session, parent/carers had the chance to engage in a forum theatre play and take the perspective of an uncle raising a teenager who is suspected of engaging in risky behaviours. The interactive performance ended with a discussion about the youth workshops and the parent/carers had the chance to vote on allowing learners to participate in the workshop. Teachers had also requested more information about the project, as the experiential learning techniques were very different from the traditional forms of instruction commonly practiced in schools.

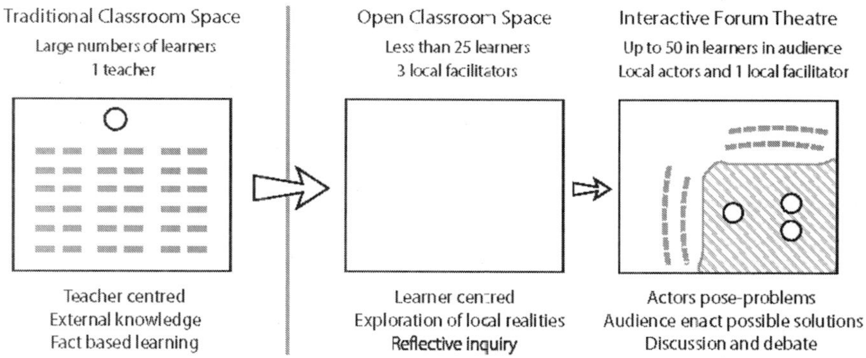

Figure 6.4. Changing the classroom configuration to enable engagement

During the school workshops, youth participated in reflective dialogue on peer influence, gender roles in relationships, and the economic and social factors around motives and expectations for intimate relationships. The school cooks who run the feeding programmes at the Zisize Schools reported seeing a change in attitude among some of the learners. This was significant as the cooks have strong relationships with the learners and, through the school feeding programme, witness the social interactions that happen outside of the classroom and on weekends.

Parent/Carer events provided an opportunity to build relationships with the community beyond the original youth target group. The Director of Zisize was present to introduce the project and bring an awareness of the organisation's work in the community. The discussion during and following the performance was an open forum to discuss issues related to youth choices and social practices. The community response was very promising with parents openly discussing issues in isiZulu language and parents requesting more discussion forums through theatre. The following quotes are translations of responses from participants in the Parent/Carer events:

> "This needs to go further into the community." (Male parent, Manyiseni Primary)
>
> "We want more drama to help us talk to our children." (Female parent, Mponshini Primary)
>
> "This needs to be taken to the high schools." (Female parent, Ekuhlehleni Primary)
>
> "I had heard of these condoms, but I had never actually seen one until today." (Male parent, Okhayeni Primary)

The depth of engagement during these parent/carer events made visible the potential for applying SDT workshop processes to engage adult community members in dialogue around priority issues.

The development of the Zisize Drama Project reflects the evolving engagement capacity that was built with the local schools and community around Ingwavuma. A call to action by local teenagers in 2004 led to training of previously unemployed adults who facilitated a pilot project, which transitioned to a multi-level community engagement for youth, parents and teachers. As well as building capacity to engage youth in reflective inquiry, other process facilitation skills were developed and extended to programming at Zisize. Parent sessions strengthened relationships with the community. Each developmental phase of the project was a response to feedback and emergent opportunities on the ground.

Peer Influence SDT Process Reaches a Mature Stage

In 2009, the Zisize Drama Project was developed further during a Health Canada funded project, *Capturing the unspoken: using applied theatre to inform programming and policy*. This project focused on developing recording frameworks and processes to capture the stories and realities shared by workshop participants in a form that could inform local programming or policy. This developed engagement capacities within the organisation by increasing skills of the facilitators and creating a mechanism to take the local knowledge shared by learners beyond the workshop space to inform programming. The Ingwavuma facilitators also travelled to townships in Eastern Cape to assist with training of project partner staff at a CBO in East London, South Africa. This included performing a forum theatre play to local residents of Duncan Village, an informal township. This created the enabling conditions for ADCID to trial an SDT workshop looking at environmental health issues in two townships in the Eastern Cape. It was this trial engagement, exploring the environmental issues among community members affected and infected by HIV/AIDS, which interested the Department of Human Settlement. They were looking to address the housing needs of residents living in formal and informal townships. More information on the *Capturing the unspoken* project is available on the ADCID website www.adcid.org.

Case Example of SDT Used for Community Development

Building Readiness for Piloting an Adapted Form of SDT

In 2012, ADCID collaborated with Tholulwazi Uzivikele to incorporate SDT workshops into a Water Access, Sanitation & Hygiene (WASH) project focused on increasing community resilience to water issues in Northern uMhibyalingana, an area covering 48 local areas, each with an induna (headman). Tholulwazi Uzivikele means "empower yourself through knowledge" in isiZulu language. TU has developed extensive technical capacities around service provision in the local area, including monitoring and evaluation supported by Geographic Information Systems (GIS) for recording of data about local homesteads. In 2012, the organisation decided to realign its focus, with each department redefining its engagement in terms of community-based outcomes instead of descriptions of activities and the services being delivered. This shift to empowerment, capacity development and community partnerships involved a shift in engagement from transferring knowledge and servicing needs to developing the "power to empower".

TU's shift toward asset-based approaches aligned closely with the development processes used by ADCID. In November 2011, ADCID engaged the entire staff of TU in an experiential visioning workshop using participatory drama. The staff experienced first-hand embodied process, where they reflected on relevant issues using experiential learning and reflection. The insights staff gained helped TU in their internal conversations towards building and deepening existing engagement with community groups, and strategies for forming new partnerships from which TU could gain community-based learning. TU management felt that Socio-Drama Topography (SDT) would be an ideal process to leverage a shift in community engagement and negotiate new relationships for empowerment and change. The trialing and piloting of the SDT process was incorporated into TU's WASH Dialogue Initiative, funded by Oxfam Australia, with a focus on facilitating deep community conversations around local water issues.

The WASH SDT pilot required initial strategy meetings to form a common yet flexible agenda for TU, ADCID and the community. ADCID took a lead around ethical and process facilitation considerations relating to SDT, TU management led on operational considerations and consulted with TU fieldworkers to help devise an implementation strategy.

Choosing Water Issues as a Theme for Exploration

The uMhibyalingana area faces many water challenges linked to systemic, environmental and social factors. For instance, there has been an introduction of eucalyptus plantations in some areas, whose trees withdraw large amounts of water from the ground causing a lowering of the water table. The impact of this technical intervention has ripple effects across different water catchment areas. Water levels have dropped which impacts the way people collect and transport water. Some communities can no longer float water containers across the lakes and must now rely on donkeys for this task, which takes significantly more time. There are also issues with contaminated water through periods of drought and following interference by cattle. Water quality in the area is so unreliable the local hospitals promote exclusive breastfeeding rather than formula milk for all mothers including those who are HIV positive. The lack of continual access to clean water for making formula milk means that the health risks of diarrhea and stomach lesions significantly outweigh the risk of mother to child transmission of HIV through exclusive feeding via breast milk alone.

Issues regarding water access are being addressed through technical solutions such as building water towers and the drilling of boreholes. However, lack of local understanding of the context can mean that over time such technical interventions are not effective and sustainable. The interconnected web of issues relating to water is complex and shifting, requiring an understanding of the local systemic, social and environmental as well as technical factors. Rapid appraisal tools such as surveys can enable external experts to look for linear cause and effect relationships for use in planning technical tasks. Unfortunately, complex social and ecological water challenges in these remote areas cannot be addressed solely with such reductive tools. More adaptive forms of engagement are necessary if the complex realities of water are to be addressed rather than single causal factors tackled in isolation. Many indirect influences affect residents' water access and quality, and the factors limiting change occur on multiple levels. The community members have first hand knowledge of this reality and its historical background. They also have indigenous knowledge of the environment and the seasonal variations in water resources. Therefore, they are an essential knowledge resource in dealing with the complex water issues they face.

The WASH SDT pilot aimed to give insights and community-based perspectives on the complex inter-relationships occurring in the social and ecological landscape around water. While quantitative data on linear cause and effect analysis of technical water solutions would be important to addressing the issues through tangible capacities, the use of SDT was seeking to uncover, harness and capacitate the intangible knowledge that is present in the community, but sadly distant from the development processes that affects the local population and consequently the provision of water related services which are so important to people's lives.

Establishing Project Organisational Capacity

ADCID and TU began by developing a common agenda to help achieve a wider understanding of the social landscape and addressing the challenges around building trusting community engagement, agreeing to joint goals, and arriving at common indicators to which the collective set of involved actors will hold themselves accountable in making progress. The capacities were scoped to enable the community mobilisation team to conduct workshops in selected local communities (isigodis) that represent each of the five water-source zones. The workshops linked access to water with sanitation and health, as well as related issues such as access of cattle to human consumption water points. The experiential workshop (SDT stages 1, 2, 3) would explore where the community came from in terms of WASH-related issues, where it found itself, and what it saw as its own assets in respect of WASH. The follow up (SDT stage 4) involves participants reflecting on the experience of building relationships among the group and the co-creation of knowledge oriented to the theme of local water access and sanitation. This deliberative stage provided a chance to synthesise the understandings made during the experiential WASH workshops (see Figure 6.5).

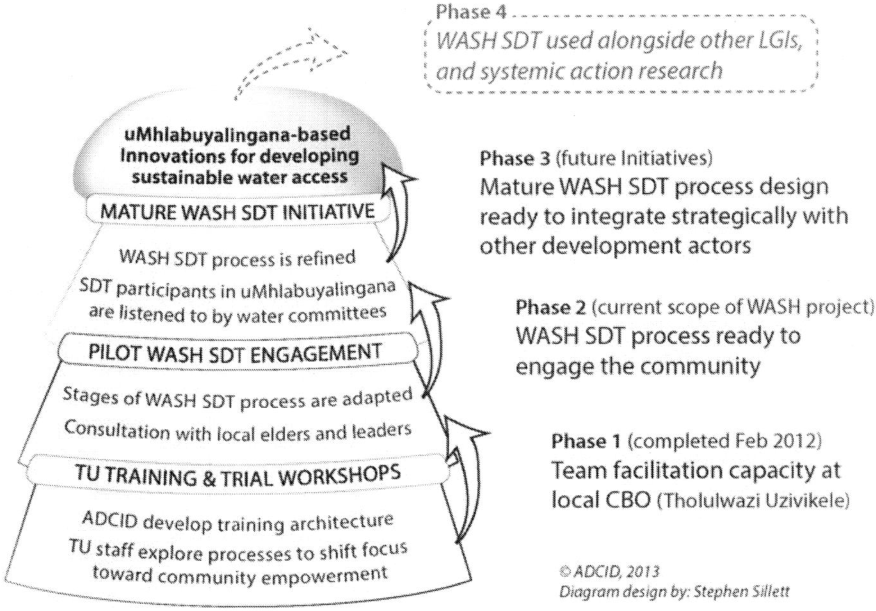

Figure 6.5. Phases of readiness and capacities for WASH SDT project

Five zones were selected in the Umkhanyakude district based on how communities were thought to obtain a significant amount of their water. The physical geography of the area influenced the five zones selected: protected wetland; unprotected wetland; along an open river; shallow groundwater; and municipal water supply.

TU's Community Mobilisation team would facilitate the SDT workshops. Some already had skills in participatory drama following training in Ingwavuma alongside Zisize Drama Facilitators in 2008 and 2009. They also had experience of counselling and community engagement through their work, which shortened the training time for facilitation. TU had a network of volunteers spread across the district providing home-based care often in remote areas. Connections were made with the local Indunas and community volunteers in the five areas who organised the attendance of local community members at the workshops. School Principals in the areas were also contacted to arrange the forum theatre presentations and other WASH related activities.[1]

Figure 6.6. Trial workshop in Manguzi developing map for WASH SDT process

A group of TU volunteers participated during three days of the staff training sessions, which gave hands-on opportunities for adapting and developing the WASH SDT process. Following the three weeks of staff training, a full-scale trial workshop located in a rural area 30km from Manguzi was implemented, with 30 TU volunteers attending from a 20km radius. Four members of TU's Community Mobilisation team and one experienced Zisize facilitator attended the three-day trial workshop. Through learnings made in this and a further trial workshop, the facilitation skills and operational capacities were deemed sufficiently developed to go beyond engaging volunteers and ready for engaging the selected communities (see Figure 6.6).

SDT WASH was designed with certain key dimensions relating to the theme, one of which was around connecting community members to the routes they take to collect water. The participants explored the primary water sources available at different times of the year and performed various stories related to experiences with water in the community. The deliberative discussion during SDT stage 4 was shaped around generating proposals for internal and external actions that could assist the community.

When engaging communities in a process that explores possibilities for social change, then ethical considerations must be made regarding community expectations. When dealing with water issues that affect participants' health and survival this is even more important. Workshop participants come on the first day with their own expectations often related to hope for assistance or an increase in services. The TU facilitators explain the nature of the WASH SDT engagement and that TU wants to learn from the community and they invite them to share their experiences and priorities around water issues through drama. After an introductory experiential exercise providing participants with a frame of reference for the kind of exploration that will occur over the three days, there is a discussion around workshop expectations. To enable reflection and Bohm dialogue, this discussion explains the open nature of exploration through the WASH SDT process and that there will be no pressure to make decisions that will be judged as definitive during the workshop. They are also informed that there will be an opportunity for analysis and proposals to be submitted to TU during a follow up session (Stage 4 of SDT).

Sharing of Stories and Knowledge

The stories shared revealed the similarities and severity of challenges experienced by the community members in all five water zones. A 55 year old woman shared her experience of spending over four hours a day fetching water–involving a thirty minute walk to and from the borehole to fill her container and a thirty minute rest, three times a day. Others found that during certain times of the year, water sources would be dry and they would have to boil dirty water, a particular problem when caring for patients who need to take anti-retroviral medication to treat HIV/AIDS. A drama scene set in a zone where rivers are a main water source revealed a man being attacked by a crocodile. Another scene showed a girl who almost drowned when she fell into a well and her lungs filled with water.

Although scenes depicted difficult issues, there were also moments of laughter in the workshop. In TU's report to OXFAM Australia for the period July-December 2012, staff reported

> As much as water is a serious and sensitive subject in most of the communities, using SDT allows people to share their hardship in a fun and humorous way. (Meyer 2012, 20)

After the WASH SDT workshops, community members attended a follow up session, which constitutes the fourth stage in this pilot SDT process. Participants were eager to discuss issues further and possible ways of responding. The session directed participants to discuss Internal Actions, which can be implemented by the community with their existing resources and capacities, and External Actions, which require support or action from outside the community.

Learnings to Inform Strategies from Moving Beyond Engagement to Community Driven Responses

During the fourth stage, community members identified various internal and external actions. The internal actions suggested by the community included educating other community members, communicating with local chiefs to take action on issues, and cooperative tasks such as coming together to clear river banks or selecting a location and pooling funds for a borehole. The external actions suggested by the community reflected a desire to increase communication with municipalities and water schemes, and advocate for increased infrastructure and fairer tariffs. Suggestions included calling on governments to intervene and assist where water schemes are charging high tariffs, calling on the municipality to fix broken taps, pipes and boreholes, and to supply reservoir water to more areas. Water committees were planned in some zones as a response to the workshops.

The WASH SDT process is presently in the pilot stage and the next step would involve deeper integration of the experiential workshop and deliberative stages. There is still further work required to mature the WASH SDT process so that it can inform strategy and innovation. A "mature" WASH SDT process would seek to strategically connect the process design to particular change objectives happening with other development actors in the community. To get a sense of how the WASH SDT project could extend into a mature version see Figure 6.7.

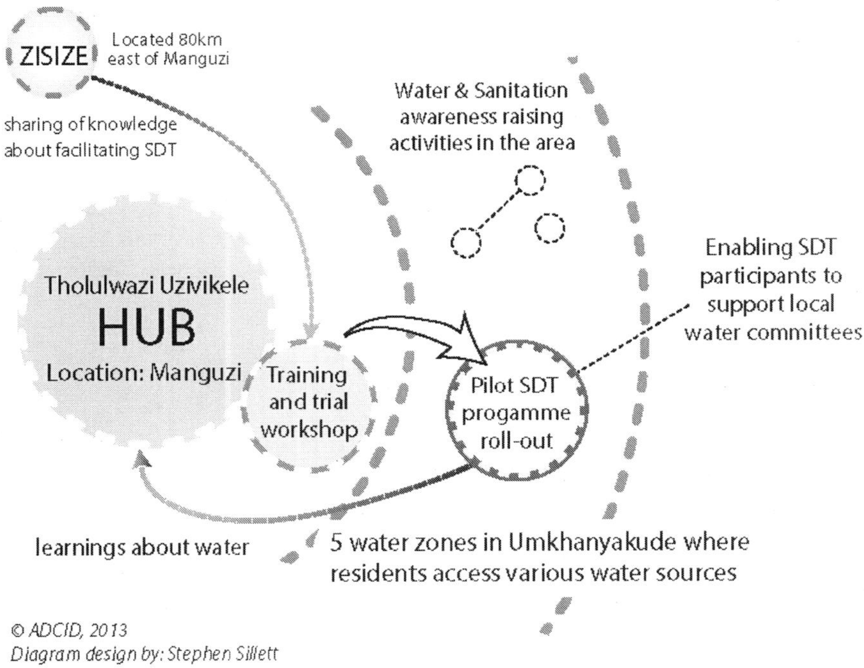

Figure 6.7. Scope of pilot phase of WASH SDT by 2013

Ongoing SDT Developments and Directions

Our most recent SDT project, *CrossGen: connecting across age and culture*, is adapting the SDT process for a very different context and theme than previous applications. The initiative, funded by the Ontario Trillium Foundation, brings together youth and adult, newcomer and long-term residents in the Niagara region of Canada to explore ways of increasing understanding and building relationships across age and culture in their communities, to facilitate a more welcoming and inclusive community. This application has completed the trial stage, which provided insights into the strategies for establishing the necessary capacities to prepare for a pilot in the future, the longer-term aim being to build sufficient capacity to inform local policy makers. This exploration proved to be productive in various ways, with emergent learning informing current and future developments of how the SDT process can enable community initiatives.

Conclusions

The adaptive development of projects and the evolving capacity yield both deep forms of engagement and innovative approaches to community development. Each developmental phase of an SDT project involves adapting and responding to feedback and emergent opportunities on the ground. This capacity focused development approach is highly contextual. The contextual and local emphasis allows marginalised communities to take part in development discourse through knowledge co-created in the SDT process.

SDT is a flexible process, but as with many large group interventions requires careful consideration before adapting to the developmental, organisational and thematic material at hand. Every adaptation involves multiple steps that help to build trust, open dialogue and allow for an in-depth exploration of issues by participants. It is a facilitation intensive process, which can be undertaken by local staff with specialised training and, for more mature versions of the SDT process, involves facilitator accreditation. While our focus here has been on utilising the SDT process for community development, SDT can also be transformative when used in a behaviour change context in schools. The case examples show community development applications working with marginalised community members as collaborators in deep rural environments. ADCID is continually exploring other contexts and applications for the process, such as multicultural urban areas in Canada.

Many change processes begin from the problem space and try to find abstract potential solutions, which are made increasingly more concrete. The SDT approach proposed by ADCID, however, starts with the concrete environment, history and social landscape and leads to enacting concrete inquiry that speaks to lived realities. It is from this grounded and open understanding of what is "out there" that the beginnings of designing solutions and selecting material to be articulated or advocated to those with access to resources occurs.

We see the SDT process as beginning a journey of exploration in terms of themes to be explored and also discoveries around the experiential and transformative aspects of the process. Regarding community development applications, we see the SDT process as contributing to the field of social innovation and large group interventions. After community members have explored the social landscape around the theme through SDT, they exhibit a remarkable level of insight and transformation in terms of expressing knowledge generated through lived experiences. This efficacy and collective wisdom changes the readiness of the community for engaging in other initiatives.

SDT is a divergent and generative process, which could subsequently lead into other convergent group exploration processes–an area that we feel deserves extensive research. A central aspect of this would be in adapting other LGIs that follow convergent and consensus based approaches for use with marginalised communities. As more communities gain the opportunity to explore creatively and deliberatively around a priority issue using mature versions of SDT, we see new forms of social innovation becoming feasible. Our proposition is that communities can develop sufficient efficacy to then engage with external experts and decision makers to sustain practical and systemic change.

Notes

1. The SDT Workshops ran alongside other WASH activities, such as Disaster Response and Relief Workshops and forum theatre performances in schools addressing water sterilisation techniques for application in house care activities, hand-washing and food hygiene. Forum theatre performances in schools engaged learners, teachers and the school cooks in discussion regarding sanitation and hygiene. After the sessions, learners volunteered to bring water, soap and buckets to school to make sure they wash their hands before and after eating food and using the toilet. Schools undertook to form WASH committees in 2013.

References

Beardon, Hannah. 2012. *Planning for uncertainty: development practice in awareness of complexity. Workshop report and reflections.* http://wiki.ikmemergent.net/files/1202-PBC_workshop_report-final.pdf

Bunker, Barbara B., and Billie T. Alban. 2006. *The handbook of large group methods: creating systemic change in organizations and communities.* San Francisco, CA: Jossey-Bass.

Chambers, Robert. 2012. *Provocations for development.* Rugby: Practical Action Publishing.

Ellinor, Linda, and Glenna Gerard. 1998. *Dialogue: rediscover the transforming power of conversation.* New York: John Wiley and Sons.

Govender, Eliza M. 2011. Development and health communications for HIV/AIDS prevention. In *Development and public health communications*, eds K. Tomaselli and C. Chasi, 51-76. Cape Town: Pearson Education.

Nicholson, Helen. 2005. *Applied drama: the gift of theatre*. Basingstoke, UK: Palgrave MacMillan.

Kaplan, Allan. 2000. Capacity building: shifting the paradigms of practice. *Development in Practice* 10(3/4): 517-526.

Meyer, Kobus. 2012. *Oxfam in South Africa program report: increasing community resilience in Northern uMhlabuyalingana*. KwaNgwanase, KwaZulu Natal.

Oxfam. 2011. *Australia Africa community engagement scheme: community led access to Water, Sanitation & Hygiene program design 2011-2016*. http://www.ausaid.gov.au/countries/sub-saharan-africa/Documents/aaces-oxfam-design.pdf.

Patton, Michael Q. 2011. *Developmental evaluation: applying complexity concepts to enhance innovation and use*. New York, NY: Guilford Press.

Purser, Ronald E., and Thomas J. Griffin. 2008. Large group interventions: whole system approaches to organizational change. In *The organizational development handbook*, ed. Tom Cummings, 261-277. Thousand Oaks: Sage.

Stats SA. 2008. *Community survey 2007: statistical release basic results municipalities*. Pretoria: Statistics South Africa.

United Nations General Assembly Special Session on HIV/AIDS (UNGASS). 2012. *Global AIDS response progress report 2012: Republic of South Africa*. http://www.unaids.org.

PART THREE:

PUBLIC MEDIATIONS OF KNOWLEDGE

CHAPTER SEVEN

NEWSPAPERS AS THE ARENA OF SCIENTIFIC CONTROVERSY: THE DEBATE ABOUT THE "ORCE MAN" IN THE SPANISH MASS MEDIA

MIQUEL CARANDELL

Introduction

The Orce Man public controversy began with the statement, on the front page of the Spanish newspaper *El País,* that the presumed "first European" hominid found the previous year in the Venta Micena site in the town of Orce was, in fact, a donkey. Following this revelation, the mass media, especially newspapers, became the scenario for a scientific debate. This controversy was a scientific controversy but rather than being played out in the usual scientific journals, it appeared in the general media, becoming a public controversy. My aim in this paper is to show how this process–science turned to media–worked in the Orce Man case and how, using the main science communication models available, this case can enhance our understanding of the role of science communication in the creation of scientific knowledge (Secord 2004, 661).

In order to achieve this main aim, I will discuss two science communication models and apply them to the Orce Man case. I will also discuss how a science in the press "genre distinction" can be applied to this controversy. Taken together, these models highlight crucial aspects of the relation between science and its communication but none of them fully considers the role of mass-media, especially newspapers, as a scenario for the circulation of scientific knowledge.

First of all, it is very important for this study to distinguish between two types of science communication in the press identified by Bucchi and

Mazzolini (2003, 7): on the one hand, "science-popularisation" included in a science section or in a special section of the newspaper and dealing mainly with scientific progress and discoveries; and on the other hand, "science as news" usually dealing with controversial issues and filling more general sections of the newspaper. Although some of the controversy counts as popularisation in the Orce Man case, both genres are clearly recognisable in my analysis.

Secondly, I will present the "expository continuum" model of science communication proposed by Shinn and Cloître (1985). This model stated, controversially, that it is impossible to distinguish clear boundaries between "pure science" and its "popularisation" (see also Hilgartner 1990). Accordingly, a "continuum" of genres of science communication must be considered wherein different categories of texts cannot be fully separated from each other. In my study, this model usefully shows how different genres of science communication mingled to form this "continuum" perfectly displayed. Moreover, I also aim to show how in the Orce Man case study, all these genres of science communication were used by scientists in their popularisation effort in order to achieve their purposes.

Thirdly, I will use the web model of science communication proposed in a seminal paper by Bruce V. Lewenstein (1995) to show that, as in the cold fusion case, science communication is understood in the Orce Man as different genres that interact with each other in very different ways to shape a web of communications. In the Orce Man case, newspapers acted as a place to present and to circulate scientific knowledge during a long period of time, not only during what Lewenstein calls the "unstable" period.

Therefore, I will conclude that the Orce Man controversy brought science to a "different dimension" where communication became crucial in the shaping of scientific knowledge. Media, especially newspapers, acted not only as a stage where science and controversy were presented to the public before discussion and publication in the scientific forums, but as a stage where new scientific knowledge was established for the wide scientific community. In addition, this will shed light on how scientific knowledge validation occurs in the mass media; it has to follow media logic. Science validation occurs not only on a different stage but also with different rules, actors and steps through "medialisation" (Weingart 1998; see also Rödder *et al.* 2012). Finally, I will attempt to show how scientists themselves were aware of this situation and acted accordingly.

Palaeoanthropology and human origins research in general are sciences that, as Bucchi (1998, 93) states in the case of cosmology, cannot promise any material or practical benefit to politicians or science policy makers in order to get further funding. However, both sciences attempt to answer profound questions about who we are or where we come from. These rather philosophical or even theological questions and their possible answers never fail to fascinate and capture the imagination of the general public.

Furthermore, palaeoanthropology is a controversy-prone science. Human fossils are difficult to find and when discovered they are often very fragmented. It is very unusual to discover an almost complete skeleton or a complete skull. Palaeoanthropologists usually work with very few, small and broken bones. Moreover, individual researchers tend to emphasise the significance of their own discoveries for human genealogy in order to get the public's favour and therefore, further funding (e.g., Hochadel 2013). This can collide with other researchers' interests and often leads to open rivalries and debates. Due to the public character of palaeoanthropological science, as the Orce Man case shows, what starts out as a circumscribed scientific skirmish can turn into a long and arduous public controversy (Lewin 1987).

Due to their prevailing role in the controversy, I have analysed three Spanish newspapers as my primary public media source: *La Vanguardia*, *El País* and *ABC*, all major and widely circulated throughout the whole country. I have also analysed science popularisation magazines–*Muy Interesante, Revista de Arqueologia* and *Investigación y Ciencia*, the Spanish version of *Scientific American*–and popular science books, Gibert (2004), Campillo (2002) and Martínez (1993). Finally, I have analysed scientific media in depth: scientific journals, self-published monographs and conference presentations.

I have divided this paper in two main parts: first, I have laid out a chronological history of the Orce Man controversy and its public presence; secondly, I have applied these science communication models to understand how this scientific dispute developed in the media.

The Orce Man Story: From "a Scientific Revolution" to the "Orce Donkey"

In the summer of 1982, Josep Gibert, Salvador Moyà-Solà and Jordi Agustí, three palaeontologists from the Institut de Paleontologia de Sabadell, none of them expert in palaeoanthropology, discovered a little cranial fragment in the Venta Micena site in the town of Orce, near Granada, Andalusia, in southern Spain. Gibert was 41 years old and was expert in fossil insectivores, Agustí was 28 and was expert in fossil rodents and Moyà-Solà, 27, had just read his PhD on fossil bovids. They named the fragment VM-0 and attributed it to *Homo sp.* in *Paleontologia i Evolució,* the Institut de Paleontologia de Sabadell journal (Gibert *et al.* 1983). Although the internal part of the bone was not visible because it still adhered to a calcareous rock, VM-0, a 10cm cranial fragment with parts of the parietal and occipital bones had some characteristics, mainly a wide cranial curvature and thinness, crucial to attributing it to a hominid. In addition to this publication, Gibert, Moyà-Solà and Agustí gave a press conference in Granada in July 1983. "The Orce Man", as headlines called it, had a great impact in Spanish media, where it was described as "a scientific revolution" (Aguilar 1983)[1]–and it really did seem a revolution. Gibert, Agustí and Moyà-Solà claimed that VM-0 was between 0.9 and 1.6 million years old. At that time, although several researches claimed an earlier arrival of hominids to Europe, the oldest acknowledged European remains were the "Tautavel Man" and the "Mauer" mandible, no more than 0.5 million years old (Gibert 2004, 273-294).[2] For these researchers, the Orce Man antedated the "oldest European" by nearly one million years. In the Spanish press, the Orce remains became the "First European" (e.g., De Semir 1983).

During the first year after this publication, the three discoverers used different media in their legitimising discourse around the Orce Man: they gave lectures (Gibert *et al.* 1984), appeared in radio and television, wrote their own newspaper and magazine articles (Agustí *et al.* 1983a, 1983b) and were even preparing a popular science book on the finding (Piñol 1984).

From October 1983 to April 1984, the internal part of the bone was cleaned in the Archaeological Museum of Barcelona. During this cleaning process an unusual crest appeared. This crest seemed to steer the evidence away from hominids and bring it close to *Equus*, the horse and donkey family genus (Campillo 2002, 31).

Gibert, Agustí and Moyà-Solà brought the fragment to Paris where the prestigious French palaeoanthropologists, Marie Antoinette and Henry de Lumley examined it. The de Lumleys were famous for overseeing several excavation sites in France and abroad. They were also the discoverers of the "Tautavel Man" and at that time, Henry was director of the Prehistory Laboratory in the Musée de l'Homme and director of the Institut de Paleontologie Humaine, both in Paris (http://fr.wikipedia.org/wiki/Henry_de_Lumley). Marie Antoinette de Lumley, expert in anatomy, compared the bone with different animal skulls and concluded that it was not a hominid bone but rather that of a young donkey (Gibert 2004, 48). Agustí and Moyà-Solà accepted the de Lumleys' verdict but Gibert held that a more detailed study of the bone was necessary. He still believed in the possibility that the fragment belonged to the genus *Homo*. Gibert left the laboratory in complete disagreement with the de Lumleys.

The next morning, on 12 May 1984, the front cover of the Spanish newspaper *El País* announced "serious evidence" that the famous Orce Man might in fact be a young donkey (Relaño 1984c). The journalist's source was clear: "Marie-Antoinette de Lumley last night in a phone conversation with this newspaper..." (Relaño 1984b).[3] The *El País* front cover touched off an acrimonious and lasting controversy. In the next four days, *El País* published more than 10 articles on the Orce Man. At that point, everybody weighed into the controversy: journalists, scientists, politicians and others (e.g., Editorial 1984; El Papus 1984; Relaño 1984a).

One year later, in May 1985, three palaeontology professors from three Spanish universities submitted a letter to the main Spanish newspapers with their opinion on the Orce bone, criticising the early popularisation of the finding and hinting that Gibert was not a serious scientist (De Renzi *et al.* 1985). According to Gibert (2004, 56-57), the palaeontology professors' statements harmed his scientific prestige as well as the credibility of "his" Orce scientific project.

Next summer, in 1986, an excavation permit for Venta Micena was denied to Gibert for the first time. In the more than 20 years of controversy, from 1986 to 2007, Gibert (2004, 105) had excavation permits in the Venta Micena site, where the Orce Man was found, only three times, even though he applied for one every summer and although his team excavated other nearby sites. In September 1986, he attended the first World Archaeological Congress in Southampton. Gibert's presence at this Congress was followed almost daily in Spanish newspapers (e.g., Moreno 1986; see Efe 1995). Although this conference was not exclusively palaeoanthropological, the press presented it as the place where the truth about the Orce Man would be debated among the scientific community (Paredes 1986). The day after Gibert's presentation, *La Vanguardia* headlined: "Gibert was not discussed in Southampton while presenting the evidence on the Orce skull" (Piñol 1985).

The following year, 1987, Gibert received an excavation permit for Venta Micena again. In October, after the excavation period, he attended another conference in Torino, Italy, which was also attended by Marie-Antoinette de Lumley. According to Gibert (2004, 78), when he presented his paper on the Orce fragment, Marie-Antoinette did not ask any questions or make any criticisms and due to her silence nobody else did either.

While Gibert was in Torino, Spanish newspapers announced the further publication of the first academic paper that followed the de Lumleys' donkey claim (Mercadé 1987). The authors of this paper, published some days later in the Spanish Journal *Estudios Geológicos*, were the Orce Man co-discoverers Agustí and Moyà-Solà (1987). They claimed that Gibert was using them in order to reinforce his Orce Man claims and they wanted to clarify their own position in the controversy. This position had to be well established in both popular and scientific media. In 1988, an excavation permit was again denied to Gibert.

To sum up, during a little over three years, from May 1984 when *El País* published the de Lumley reclassification, to October 1987 when Agustí and Moyà-Solà published the first scientific paper dealing with this attribution, the supposed "scientific knowledge" that the bone VM-0 was a member of the horse family was discussed in the press. The scientific discussion did not take place in scientific forums such as the conferences at Southampton and Torino, at least according to Gibert. Thus, during these three years, the stage for the scientific controversy was exclusively public.

Orce Human Palaeontology Conference: From "the Oldest European" to "a Common and Ordinary Horse"

The following years saw less media attention while Gibert and his team continued their in-depth study of the bone fragment. During these years, Gibert also claimed that more than a million years ago, hominids crossed the Gibraltar strait, a claim that was not accepted in large sectors of the Spanish scientific community (Gibert *et al.* 2003; for an opposing view, see Carbonell and Estévez 1983, 68). Some of the team members were: Bienvenido Martínez, Gibert's former doctoral student and Orce's animal fossils expert; Paul Palmqvist, biologist and mathematician who had done a fractal mathematical study of the Orce bone; and Lluís Gibert, Josep's son and an expert in geology.[4] With this team, Gibert published several papers in all kinds of scientific media. Less than 10% of these articles appeared in citation index journals. The other articles were published in self-published monographs, conference proceedings or second-line scientific journals.[5]

In 1995, after more than ten years of controversy, Gibert managed to hold an International Human Palaeontology Conference in Orce. With this conference, media attention reappeared.[6] In his popular science book, Gibert (2004, 91) explains that during the conference, journalists attended the sessions, asked questions to scientists and visited palaeontological sites. At the Conference, Gibert (1995) and his team presented several lines of evidence for the acceptance of the Orce Man. According to the press, the attending experts accepted this evidence. During the Conference, *La Vanguardia*, another major Spanish newspaper, published an article with the title "An international conference ensures that the Orce Man is the oldest European" (Efe 1995; see also Turbón 1995). In the closing session of the Conference, Diego Valderas, Andalusian Parliament president, promised funding and excavation permits to Gibert and finally stated that "Gibert had won the public opinion battle" (F.R.A. 1995). This "public victory" did not last too long. Next month, *El País* published statements on the Orce bone by José María Bermúdez de Castro and Eudald Carbonell, both directors of the well-known Spanish site Atapuerca (Hochadel 2013). Carbonell criticised Gibert's scientific rigour and Bermúdez de Castro, who attended the Orce Conference, stated that "according to the fossils, there is no Orce Man" (Rivera 1995).[7]

Next summer, 1996, two of Gibert's former collaborators, Bienvenido Martínez and Paul Palmqvist, together with Alain Turq, a French archaeologist who had studied stone tools from another site near Venta Micena, announced first to the local and then to the national press their split from Gibert's team. They argued that they were tired of Gibert's lack of funding and his "obsession" with the Orce fragment. Palmqvist also accused Gibert of fraud. According to Palmqvist, Gibert gave him a simplified sketch of the bone, which altered the results of his fractal analysis (De León-Sotelo 1996; Editorial 1996).

Almost a year later, Moyà-Solà and Palmqvist published two scientific articles regarding the Orce bone in the June 1997 issue of the international *Journal of Human Evolution*. In his article, Palmqvist (1997) revised his fractal analysis with a new Orce bone sketch: the bone was no longer human and was closer to the horse family. Moyà-Solà's paper, entitled "The Orce skull: anatomy of a mistake", addressed VM-0 anatomy and concluded that VM-0 was "just a common and ordinary horse" (Moyà-Solà and Köhler 1997). The publication of these papers was reported in Spanish newspapers, which noted that this new "demolishing evidence" was published in a prestigious and international journal (Editorial 1997). Both *El País* and *La Vanguardia* reproduced exactly the same sentence: the Orce Man is "a common and ordinary horse" (Ruiz de Elvira 1997).

In 2007, after years of less media attention but an ongoing controversy over the Orce bone and the Venta Micena excavation permits, Josep Gibert i Clos died due to a fulminating cancer. His ashes were scattered in the Venta Micena site (Planas 2007). Although the Orce Man regularly appeared in Spanish magazines and newspapers, with Gibert, the Orce Man's main defender, the controversy itself also faded away.

Science in the Press: Popularisation or News?

During the first year after their finding, Gibert, Agustí and Moyà-Solà published a series of full-colour articles in *La Vanguardia*'s Sunday magazine that described the characteristics of the bone and the site, the scientific conditions in which it was found (not very good because of a lack of money) and the Orce Man's prominent position in the "humanisation process" (Agustí, Gibert and Moyà-Solà 1983a). Some years later in a period of less media attention, Gibert (1993), this time alone, published in the same newspaper's "Science and technology" section, an article explaining the major achievements of his research group in spite of the lack of funding and excavation permits. Gibert also established the "humankind" of the Orce bone fragment and its enormous significance as the "first European". Both articles were clear examples of what Bucchi and Mazzolini (2003, 7) call "science-popularisation" in the press: first, a new, exciting discovery presented by the researchers themselves in the Sunday magazine; secondly, a non-controversial article, the controversy only present by way of introducing the fragment's history and with several "definitive" lines of evidence in favour of the hominid attribution, written again by the researcher and in the "Science and technology" section.

However, the dispute explosion on the *El País* front cover, the following days with statements from the scientists involved, the press attention to both the Southampton conference and the Orce conference, and the following statements from Carbonell and Bermúdez de Castro, are all clear examples of what Bucchi and Mazzolini (2003, 7) call "science as news". In these cases, science is on the front cover of the newspaper and dealing with harsh controversial issues. Thus, science appears clearly "open" in a way that allows us a better understanding of how it works inside the scientific community and in relation to different "actors" as politicians (Latour 1997).

Together, these examples suggest the suitability of the "two distinct journalistic genres" of Bucchi and Mazzolini's analysis in the treatment of the Orce Man controversy in Spanish newspapers. But they show us more than just this "genre distinction". The first examples of popularisation of the Orce Man both before and during the controversy indicate how scientists used media, in this case newspapers, in order to get public, journalistic and institutional favour. In one of the first series of articles with the subheading "Lack of funding", Agustí, Gibert and Moyà-Solà (1983a) when talking about Spanish palaeontology in general were quite explicit in their quest for support: "Institutions should provide a minimum of resources [to researchers]". Finally, on their commitment as the Orce Man discoverers, they added: "with the necessary institutional support... we hope to fulfil our responsibility".[8] Thus, together with a long exposition of the palaeontological work on the Orce Man characteristics and its significance, Spanish researchers presented their work in the press in order to attract politicians' attention in a bid to secure funding for future research. As Greg Myers (2003, 270) states, "being able to explain one's project and its relevance to wider society…is an essential part of running a large lab and getting further funding" (see also Kjærgaard 2012). As I have already pointed out, palaeoanthropology does not have any direct "relevance" in society but–and as this example also shows–it can try to answer more philosophical questions such as the "humanisation process" in order to acquire this "relevance" and thence, further funding.

The second science-popularisation example shows how often this kind of popularisation in the press also tries to "wash" the scientific reputations of "marginalised" scientists such as Gibert. In his article, Gibert (1993, 10) quoted several of his scientific publications and self-published monographs, highlighting the work capacity of his scientific team seeking to recover lost prestige in the public forum during the years of controversy.

To sum up, "science as popularisation" in the press in the Orce Man case had clearly more purposes than disseminating already created scientific knowledge to the lay public. These examples show that funding purposes can shape how science appears in the press and so must be taken into account when analysing science popularisation.

Expository Continuum: Popularisation or Science?

The "expository continuum" model of science communication considers that science communication is a "continuum" of different categories of texts that cannot be separated from one another (Shinn and Cloître 1985, 58). Thus, popularisation cannot be fully separated from "true" scientific knowledge. As Stephen Hilgartner (1990, 528) has shown, the boundaries between science and popularisation are permeable: the difference between what counts as popularisation and what counts as science is "a matter of degree" (see also van Dijck 2008). Moreover, the developers of this model further claimed that a scientist's "expository practices" can also "play an active role in the knowledge production process" (Shinn and Cloître 1985, 32; see also Clemens 1986). Therefore, popularisation can be a means of validating and even creating scientific knowledge. As Richard Whitley (1985, 12) put it: "popularisation cannot be separated from knowledge generation and development but needs to be considered as part of the overall process of intellectual change".

Going back again to just two months before the public controversy and one month before the end of the bone cleaning process, in March 1984, *La Vanguardia,* again in its Sunday coloured magazine, published a long report on the bone fragment and on the Venta Micena site (González 1984). Although this time the article was not signed by the researchers but by a journalist, it quoted both Gibert's and Agustí's statements. In this article, the duo made some claims that had not appeared in their first academic article. Their new claims were that Venta Micena site was, contrary to what they had previously claimed, a long-stay human site. Animal bones found there had a distribution pattern that, according to the researchers, indicated human agency. With this claim, a 10cm cranial bone was converted into a more than 1 million year old human settlement in Europe.

A few days after the first *El País* front page with the de Lumley reclassification, Gibert presented to the press the anatomical reasons for defending the Orce Man's humanity (Arroyo 1984):

> [The crest] in our cranium is simple, in horse's is double. .... The differences in the crest are the following: It is double in the IPH (L'Institut de Paléontologie Humaine in Paris) [horse] specimen, simple in the young [donkey] specimen from the LAC (Comparative Anatomy Laboratory), but in both cases, it has a shape that enlarges at the base, never the laminar shape of the Orce Man.[9]

The same "scientific" sentence is found in a conference proceeding in which Gibert (1984, 42) published his work one month later.

These two examples show how scientists' claims can appear, sometimes in almost the same form, in both general media and scientific publications. Both examples, one of "science-popularisation" in the press and the other more closely related to "science as news", indicate then that the creation of scientific facts is not separated from its popularisation and popular media appearances. With these media appearances, the scientists wanted to reach not only the general public but also colleagues, making them "epistemologically active" public appearances. The first example shows how scientists used popular media to present their claims before publishing them in a scientific paper. The later controversy did not allow the discoverers to publish their scientific papers on the site, but this case of popularisation indicates the intention of presentation of further claims that would have appeared in subsequent scientific publications. As Bucchi (1996) has pointed out, researchers typically use general media to strengthen a claim that could be controversial in the scientific media.

Popularisation is a test field, a place for speculation, a place to release "difficult to prove" scientific hypotheses before releasing them in the scientific media.[10]

The second example shows, on the one hand, that in this case, the notion of "science as news" also carries a more "scientific" style similar to an academic publication. On the other hand, it also shows how during the controversy, Josep Gibert used all kinds of media, from the press to peer-review scientific journals, from popular science books to self-published monographs and even children's books (Gibert 2007) to defend his controversial claims across the wide range of publics and media. Gibert disseminated his ideas using this entire "continuum" of media in order to reach as many publics as possible (see also De Bont 2003, 607; Fahnestock 1989). In addition, Gibert, excluded from the general scientific communities where his claims were neither widely published nor accepted, used popularisation in order to reach colleagues and other scientists. Gibert then changed from one communication medium to another becoming scientist or populariser depending on the situation (cf. Gregory 2003, 26 and 33).

Moreover, as already shown, all through the controversy several examples shed light on how scientists from both "sides" of the dispute appeared in the media looking not only for popularisation or for public and institutional favour, but also for validation of scientific knowledge–knowledge that would, either previously or subsequently, be published in scientific publications. One example is the appearance in Spanish newspapers in 1997 of the publication of Moyà-Solà's and Palmqvist's articles in *Journal of Human Evolution*, a world leading scientific journal in palaeoanthropology (Gundling 2010, 318-319). Although this journal reached most of the specialists in the field, the public presence was also necessary for the acceptance of this knowledge. General media, in this case Spanish newspapers, had then a role in the "final rejection" of the Orce Man.[11] The popularisation or the media visibility of this scientific article also had authority in the validation of scientific knowledge–as van Dijck (2008, 390) states, "popularisation itself can become a scientific authority".

Web of Science Communication: From Media to Science?

In Bruce V. Lewenstein's web of science communication model, different kinds of scientific and popular media interact with each other in many different ways. For Lewenstein (1995, 425-428), the notion of a "continuum" is insufficient to understand science communication since it still distinguishes popularisation of science on one side of the "continuum" and "proper" science on the other. In the web model, science communication is not considered as something different or external to science, but an intrinsic part of scientific activity (see also Gregory 2003). Thus, this model uses a web of media where information goes back and forth from one medium to another with no clear distinction between science and its communication. Lewenstein (1995, 428-431) also distinguishes a first unstable period in which media is crucial from a period of more information stability when peer-review articles appear.

In May 1986, the Spanish popular science magazine *Muy Interesante*, published an interview with the scientist Maria Teresa Alberdi regarding her palaeontology work. Alberdi was a palaeontologist from Madrid who at that time was working on the geology and palaeontology of the Orce area (e.g., Alberdi and Bonadonna 1989). After some general questions about palaeontology, the interviewer asked Alberdi her opinion on the Orce bone: "I haven't seen the bone, neither has Emiliano Aguirre, number one Spanish palaeoanthropologist…[but] according to what I have seen in the newspapers..." (Torreiglesias 1986, 130).[12]

Scientists, even specialists on palaeoanthropology or experts working on palaeontology on the Orce area who did not have direct access to the bone, used Spanish media, especially newspapers, in order to obtain information on a scientific object, the Orce Man. Scientific knowledge travelled this time from a general medium to the scientists who gathered information on the bone, at least from the controversial de Lumley claim and the following debates in Spanish newspapers. As already seen, at that moment, Gibert also used newspapers to disseminate his opinions on the bone's anatomy (e.g., Arroyo 1984). Alberdi's statements show that these opinions reached colleagues through the mass media. As previously said, this period of getting information from the media lasted at least three years until the Moyà-Solà and Agustí publication.

As also already stated, the International Prehistory Conference held in Orce, was widely followed in Spanish newspapers and television. In their coverage of the conference, newspapers included several scientists' opinions on the fragment and on the Venta Micena site, and journalists also reviewed some of the sessions. According to *El País*, in the opening session the prestigious scientists Philip V. Tobias and F. Clark Howell stated they were "convinced that the Orce skull belongs to a hominid" (Arias 1995).[13] Also *El País* highlighted Paul Sondaar's presentation at the conference. Sondaar was a Dutch researcher working in Indonesia who claimed that 0.7 million years ago, hominids crossed a 20km sea gap to reach Flores Island. According to the journalist, this could support Gibert's hypothesis of Orce hominids crossing the Strait of Gibraltar (Cervera 1995). Information on the conference appeared in Spanish newspapers where scientists gave their opinion on the bone and on the site, and which became the space where a scientific forum was presented. Moreover, the Orce Conference organising committee, mainly Gibert's team, set up a big screen in Orce's main square to allow the town's inhabitants to follow the conference's scientific debates. A month after the conference, two scientists that were against the Orce Man, Carbonell and Bermúdez de Castro, used the newspaper *El País* to establish their scientific position in the controversy.

Summarising, the Orce Man case has plenty of examples of how scientific information travels from one medium to another, from what is supposed to be a scientific forum to a popular one. Scientific knowledge can "jump" from one medium to another in the supposed "continuum" creating then a "web" of media with multiple level interactions (Bucchi 1998, 21). Knowledge is not created in one space or forum but is generated, shared, circulated and validated through all the different nodes of this science communication web. Unlike the cold fusion case, in the Orce Man controversy the period in which the media was the main information source for experts and lay people was not a few weeks or months, but at least three years from the de Lumleys' statement in *El País* to the first article dealing with the donkey attribution. Lewenstein's "unstable" period, then, is not applicable to this case; rather, the mass media were also a stable knowledge circulation medium.

Conclusions: Newspapers as the Arena for Scientific Debate

As already noted, in September 1987, Agustí and Moyà-Solà announced to the press the further publication of the first scientific paper following the de Lumleys' statement. This "science per press conference" was another important node in the Orce Man's web of science communication and was widely followed by Spanish newspapers (Editorial 1987; Mercadé 1987). Two of these newspapers, *La Vanguardia* and *El País*, quoted Agustí's statement that the "debate should have been limited to scientific publications, but the media treatment of the issue has given it a different dimension" (Mercadé 1987).[14]

In my opinion, this first public appearance of Moyà-Solà and Agustí is crucial in understanding the relationship between science and public in the Orce Man case. Since the discovery of the Orce fragment, the only scientific paper that referred to VM-0 was the initial publication of the three discoverers in *Paleontologia i Evolució*. During this time Gibert attended conferences and published in proceedings but as has been stated, these conferences did not allow a scientific debate. Moreover, these papers all presented his views of the controversy. Neither Henry nor Marie-Antoinette de Lumley published any scientific paper dealing with the Orce finding. Only when Agustí and Moyà-Solà were tired of Gibert's use of their name in his claims did they decide to publish a scientific paper claiming the fragment belonged to a young donkey.

As has been shown in this study, at that moment, the scientific publication was not enough to validate a new scientific knowledge. Due to media attention, a public presentation of this claim was also necessary and, as sociologist Peter Weingart (1998, 871) has stated, the stronger the dependence of science on public approval, the more attention given to the media. Thus, a loop is created in which dependence of science on resources leads to science public appearances which lead to a necessity of public acceptance in the media and to more media attention. Science is then presented in different ways, by different rules, criteria and logic. As Weingart (1998, 872) also states, science is "medialisated" (see also Rödder *et al.* 2012, 4-5).

Over at least three years, the de Lumley reclassification in *El País* was received and accepted or criticised throughout the Spanish scientific community. Alberdi's interview is just an example of how scientists received information from Spanish newspapers. As Hilgartner (1990, 529) says, "popularisation" is a category not necessarily established by its genre but also by the context of reception (see also Gregory 2003, 26). When *El País* published the de Lumley donkey claim it acted as a scientific publication, because along with the general public, experts also received this information through the newspaper. The knowledge presented that day was not only a case of Bucchi and Mazzolini's "science as news" to be forgotten a few days later, nor just Lewenstein's first unstable period, but it was also knowledge that spread across the Spanish scientific community which already knew the bone due to the early discoverers' popularisation effort. For three years, the scientific community only knew about the donkey reclassification from Spanish popular media. The Orce Man case also shows that this kind of science media appearance can entail a more "scientific" style which would favour this reception among the expert community. Moreover, different kinds of public, such as politicians, journalists and scientists from other disciplines also received information on the Orce controversy exclusively from Spanish mass media–and more importantly, this wide range of groups also gave their opinion regarding this scientific issue in the newspapers.

In my view, all these things–the *El País* role as a "scientific medium", the necessity of the public appearance of a scientific claim and the new range of science publics–is what Agustí was talking about when he said that the Orce controversy was in a "different dimension". And probably he knew what he was talking about. During the first years of controversy, Agustí collected several newspaper clippings which I was recently able to examine. In his popular science book, Gibert (2004, 405-410) also analyses several newspaper articles in order to develop his claim that during all the controversy he has "remained in the scientific domain" and to blame "the immaturity of Spanish science"[15] for the public debate. As Marianne Sommer (2006, 209) has pointed out in her work on the early 20th century French palaeontologist, Marcellin Boule's collection of newspaper clippings on his description of the Chapelle-aux-Saints Neanderthal: "the fact that Boule collected the articles…points to a back-and-forth communication between the scientist and the publics". Information went then from scientists to newspapers and from newspapers back to scientists.

To sum up, all these science communication models allow for a richer picture of how the Orce Man public controversy developed. The early popularisation articles of the three discoverers or the *El País* account of the de Lumley reclassification show the adequacy of the notions of "science as news" and of "science as popularisation" in the press. Several examples in this study have shown how boundaries between science communication genres are difficult to establish (Hilgartner 1990, 524). Moreover, experts' use of the media also sheds light on how interactions between different media worked and how information went from one medium to another and back in very complex and different ways (Lewenstein 1995, 427-428). As I have suggested, this case study shows how in public scientific controversies, knowledge must be validated and re-established several times. This validation and re-establishment must occur in both the scientific forum, as in a peer-reviewed paper or a conference presentation and in the public forum, as in a newspaper article covering the publication of a scientific paper or a conference presentation. Public validation then does not replace science validation but supplements it (Weingart 1998, 870).

Finally, if Gibert had not insisted on the Orce Man hominid attribution, the *El País* report should have been enough to establish among the scientific community that the Orce Man was a donkey. Therefore, this report of the de Lumleys' claim was not ephemeral knowledge, but rather it established scientific knowledge in the Spanish scientific community for at least the three years prior to the Agustí and Moyà-Solà scientific publication. Hence in this "different dimension" of science, of which scientists were aware, scientific knowledge is created, legitimated and established during its communication not only in scientific media but also in newspapers and other mass media.

Notes

1. Díaz-Rojo (2011) has analysed the media rhetoric used in these first news reports of the discovery.
2. For researchers that claimed an earlier arrival to Europe, see, e.g., Carbonell and Estévez (1983, 151); De Lumley (2010 [2007], 116).
3. "Marie-Antoinette de Lumley en conversación telefónica con este periódico...."
4. The "fractal dimensions" elucidated through complicated mathematical formulae were used in this case to establish the complexity of the sutures. The simpler the suture, the more likely the fragment is within the genus *Homo* (Gibert and Palmqvist 1995).
5. According to Gibert himself (Gibert and Gibert 2000), by 2000 he had published more than 130 articles, 16 in Citation Index Journals, 11 of which were in international journals.
6. In the days before, during and after the conference, *La Vanguardia* published four articles on it, *El País* five, and *ABC* also five. Examples include De León-Sotelo (1995) and Efe (1995). The conference even attracted some attention abroad (Denison 1995; Weimer 1995).
7. "A partir de los fósiles presentados, el hombre de Orce no existe" (Rivera 1995).
8. "Las instituciones deben dotarles con un mínimo de medios" and "con el apoyo institucional necesario…esperamos cumplir con nuestro cometido".
9. "[La cresta] en nuestro cráneo es sencilla, los caballos la tienen doble. Las diferencias de la cresta son las siguientes: Es doble en el ejemplar del IPH es sencilla en un ejemplar joven del LAC (Laboratorio de Anatomía Comparada), pero, en ambos casos, presenta una forma que se agranda hacia la base, nunca la forma laminar del hombre de Orce."
10. Hochadel (2009) and Gregory (2003) have shown respectively how popular science books and science fiction can work in the same way.
11. Whitley (1985, 11) states that expository practices affect "what comes to constitute knowledge in that field at that time".
12. "Yo no lo he visto, y tampoco Emiliano Aguirre, número uno español en paleontología humana…[pero] a juzgar por lo que ha salido en la prensa...."
13. Tobias was, for example, one of the describers of *Homo habilis* with Louis Leakey. Howell is also a major palaeoanthropologist from the University of California. "Convencidos de que el cráneo de Orce pertenece a un homínido" (Arias 1995).
14. "[E]l debate debería haberse mantenido en los cauces de las publicaciones científicas pero que el tratamiento periodístico del tema le ha dado una dimensión distinta."
15. "Mantenido en el terreno científico."

Acknowledgments

Early versions of this paper were presented at the Science and the Public 6th Annual Conference (Kingston upon Thames, 2-3 July 2011), at the British Society for the History of Science Postgraduate Conference (Warwick, 6 January 2012), and at the Public Communication of Science and Technology, PCST (Florence, April 2012). I am grateful to all the participants at these sessions for providing me very interesting feedback. I would also like to thank the editors of this volume, all the history of science people in Barcelona, Oliver Hochadel for his intellectual guidance, Jesse Olszynko-Gryn for his careful reading of a draft version of this paper, my parents for their very helpful comments and constant support and Clara Florensa for her encouragement and stimulating conversation.

References

Aguilar, José. 1983. El hallazgo del Hombre de Orce puede suponer una revolución en estudio de la especie humana. *El País*, June 14.

Agustí, Jordi, Josep Gibert and Salvador Moyà-Solà. 1983a. Características del Hombre de Orce. *La Vanguardia*, August 21: 47.

—. 1983b. El Hombre de Orce. *Revista de Arqueologia* 29: 16-21.

Agustí, Jordi, and Salvador Moyà-Solà. 1987. Sobre la identidad del fragmento craneal atribuido a *Homo* sp. de Venta Micena (Orce Granada). *Estudios Geológicos* 42: 538–443.

Alberdi, Maria Teresa, and F.P. Bonadonna. 1989. Geología y paleontología de la cuenca de Guadix-Baza. *Trabajos sobre Neógeno-Cuaternario* 11: 1-355.

Arias, Jesús. 1995. Expertos creen que los fósiles de Orce son anteriores a los de Atapuerca. *El País*, September 5.

Arroyo, Francesc, 1984. Josep Gibert considera que la hipótesis mas fidedigna sigue siendo la del homínido. *El País*, May 13.

Bucchi, Massimiano. 1996. When scientists turn to the public: alternative routes in science communication. *Public Understanding of Science* 5: 375-394.

—. 1998. *Science and the media*. London: Routledge.

Bucchi, Massimiano, and Renato Mazzolini. 2003. Big science, little news: science coverage in the Italian daily press, 1946-1997. *Public Understanding of Science* 12: 7-24.

Campillo, Domènec. 2002. *El cráneo infantil de Orce: el homínido más antiguo de Eurasia*. Bellaterra: Edicions Bellaterra.

Carbonell, Eudald, and Jordi Estévez. 1983. Les primeres occupacions

humanes en els Països Catalans. *Revista Avenç* 57: 67-70.
Cervera, José. 1995. La difícil ruta desde África. *El País*, September.
Clemens, Elisabeth. 1986. Of asteroids and dinosaurs: the role of the press in the shaping of scientific debate. *Social Studies of Science* 16: 421-456.
De Bont, Raf. 2003. The creation of prehistoric man: Aimé Rutot and the eolith controversy, 1900–1920. *Isis*, 94(4): 604-630.
De León-Sotelo, Trinidad. 1995. José Gibert: "El congreso de Orce cambiará la prehistoria en Europa". *ABC*, September 2: 53.
—. 1996. Orce, una polémica que no fosiliza. *ABC*, August 8: 47.
De Lumley, H. 2010[2007]. *La gran aventura de los primeros europeo.* Barcelona: Metatemas, Tusquets Editores.
De Renzi, Miquel, Jaume Porta and Jaume Truyols. 1985. Nota sobre paleontología. *Avui*, March 6: 10.
De Semir, Vladimir. 1983. Tres paleontólogos catalanes descubren el hombre más antiguo de Europa y Asia. *La Vanguardia*, June 12: 51.
Denison, Simon. 1995. 1.8 million-year-old human presence claimed in Spain. *British Archaeology* 7: 4.
Díaz Rojo, José Antonio. 2011. Retórica científica en la prensa. El hallazgo paleontológico del cráneo de Orce (1983). In *La circulación del saber científico en los siglos XIX y XX,* ed. José Antonio Díaz Rojo, 99-128. València: Instituto de Historia de la Medicina y de la Ciencia López-Piñero.
Editorial. 1984. Hombre o borrico. *El País*, May 15.
—. 1987. Dos investigadores del cráneo de Orce afirman que perteneció a un équido joven. *El País*, October 2.
—. 1996. Polémica en Orce. *ABC*, August 11: 49.
—. 1997. Los científicos hallan pruebas demoledoras contra la autenticidad del hombre de Orce. *La Vanguardia*, August 18: 20.
Efe. 1995. Un congreso internacional asegura que el Hombre de Orce es el europeo más antiguo. *La Vanguardia*, September 5: 36.
El Papus. 1984. El misterio del Hombre de Orce. *El Papus*, June: 525.
Fahnestock, Jeanne. 1989. Arguing in different forums: the Bering crossover controversy. *Science, Technology, & Human Values* 14(1): 26-42.
F.R.A. 1995. La consejería promete apoyo económico para el yacimiento de Orce. *ABC*, September 9: 39.
Gibert, Josep. 1984. El yacimiento de Venta Micena (Orce, Granada). Su importancia, acción antropogénica y características paleoantropolígicas del fragmento de cràneo de *Homo* sp. In *Homenaje a Luis Siret*, 37-48. Sevilla: Junta de Andalucía.

—. 1993. El primer ser humano de Europa. *La Vanguardia*, May 1: 10-11.
—. 1995. "Human remains in Orce and Cueva Victoria." Paper presented at the International Conference on Human Palaeontology, Orce, Spain. September 5: 17-18.
—. 2004. *El Hombre de Orce, los hominidos que llegaron del Sur.* Córdoba: Almuzara.
—. 2007. *¿Quieres ser paleontologo?* Barcelona: EDBE.
Gibert, Josep, and Lluís Gibert. 2000. Las investigaciones de Orce y los medios de comunicación. In *II simposio latino sobre geología, medio ambiente y sociedad*, eds. Josep Maria Mata i Perelló and Isidre Roma, 227-232. Cubells.
Gibert, Josep, and Paul Palmqvist. 1995. Fractal analysis of the Orce skull sutures. *Journal of Human Evolution* 28: 561–575.
Gibert, Josep, Jordi Agustí and Salvador Moyà-Solà. 1983. Presencia de *Homo* sp. en el yacimiento del Pleistoceno inferior de Venta Micena (Orce, Granada). *Paleontologia i Evolucio*, Special Issue: 1-9.
—. 1984. Día a Día. *La Vanguardia*, February 22: 53.
Gibert, J., L. Gibert, E. Bulygina and A. Iglesias. 2003. The Straits of Gibraltar, a Plio-Pleistocene door of Europe? *Human Evolution* 18(3-4): 147-160.
González, Evaristo. 1984. El Hombre de Orce sigue asombrando. *La Vanguardia*, March 11: 49.
Gregory, Jane. 2003. The popularisation and excommunication of Fred Hoyle's "life-from-space" theory. *Public Understanding of Science* 12: 25-46.
Gundling, Tom. 2010. Human origins studies: a historical perspective. *Evolution: Education and Outreach* 3: 318-319.
Hilgartner, Stephen. 1990. The dominant view of popularisation: conceptual problems, political uses. *Social Studies of Science* 20(3): 519-539.
Hochadel, Oliver. 2009. Atapuerca–the making of a magic mountain. Popular science books and human-origins-research in contemporary Spain. In *Communicating science in 20th century Europe: a survey on research and comparative perspectives*, ed. Arne Schirrmacher, 149-163. Berlin: Max-Planck Institute for the History of Science.
—. 2013. *El mito de Atapuerca*. Bellaterra: Edicions UAB.
Kjærgaard, Peter. 2012. The fossil trade: paying a price for human origins. *Isis* 103(2): 340-355.
Latour, Bruno. 1997. *Science in action.* Cambridge, M.A: Harvard University Press.
Lewenstein, Bruce. 1995. From fax to facts: communication in the Cold

Fusion saga. *Social Studies of Science* 25(3): 403-436.

Lewin, Roger. 1987. *Bones of contention*. Chicago: University of Chicago Press.

Martínez, Bienvenido. 1993. *El Hombre de Orce. Historia de un descubrimiento* Granada: Proyecto Sur.

Mercadé, Josep. 1987. Los colaboradores de Gibert dicen que el cráneo de Orce pertenece a un equino. *La Vanguardia*, October 2: 40.

Moreno, Manuel. 1986. El cráneo de Orce es humano, según confirma el Congreso Arqueológico Mundial. *El País*, September 9.

Moyà-Solà, Salvador, and Meike Köhler. 1997. The Orce skull: anatomy of a mistake. *Journal of Human Evolution* 33: 91–97.

Myers, Greg. 2003. Discourse studies of scientific popularisation: questioning the boundaries. *Discourse Studies* 5(2): 265-279.

Palmqvist, Paul. 1997. A critical re-evaluation of the evidence for the presence of hominids in lower Pleistocene times at Venta Micena, Southern Spain. *Journal of Human Evolution* 33: 83-89.

Paredes, José Antonio. 1986. Gibert: las dudas sobre el "hombre de Orce" acabarán en Southampton. *La Vanguardia*, August 23: 20.

Planas, Mónica. 2007. Un hombre marcado por un fósil. *La Vanguardia*, October 9: 37.

Piñol, Rosa Maria. 1984. La nova editorial catalana Empúries publicará diverses obres de Riba. *La Vanguardia*, March 15: 43.

—. 1985. Gibert no fue discutido al presentar las pruebas sobre el cráneo de Orce. *La Vanguardia*, September 3: 28.

Relaño, Alfredo. 1984a. La Junta de Andalucía valora con cautela las dudas sobre el origen del "hombre de Orce". *El País*, May 13.

—. 1984b. Serias sospechas de que el "hombre de Orce" no es un hombre, sino que podría ser un asno de dos a cuatro meses de edad. *El País*, May 12: 22.

—. 1984c. Serios indicios de que el cráneo del "hombre de Orce" pertenece a un asno. *El País*, May 12: 1.

Rivera, Alícia. 1995. Paleontólogos y geólogos cuestionan la existencia del "hombre de Orce". *El País*, October 6.

Rödder, Simone, Martina Franzen and Peter Weingart. 2012. Exploring the impact of science communication on scientific knowledge production: an introduction. In *The sciences' media connection–public communication and its repercussions*, eds. Simone Rödder, Martina Franzen and Peter Weingart, 3-16. Dordrecht: Springer.

Ruiz de Elvira, Malen. 1997. El fósil del "hombre de Orce" pertenece en realidad a un caballo, según los científicos. *El País*, August 17.

Secord, James A. 2004. Knowledge in transit. *Isis* 95: 654-672.

Shinn, Terry, and Michel Cloître. 1985. Expository practice: social, cognitive and epistemological linkage. In: *Expository science: forms and functions of popularisation*, eds. Terry Shinn and Richard Whitley, 31-60. Dordrecht: Reidel.

Sommer, Marianne. 2006. Mirror, mirror on the wall: Neanderthal as image and "distortion" in early 20th century French science and press. *Social Studies of Science* 36(2): 207-240.

Torreiglesias, Manuel. 1986. Los fósiles son patrimonio de la humanidad. *Muy Interesante* 60: 129-131.

Turbón, Daniel. 1995. Hombre de Orce: Josep Gibert tenia razón. *La Vanguardia*, September 17: 64.

Van Dijck, Maarten. 2008. From science to popularisation and back–the science and journalism of the Belgian economist Gustave de Molinari. *Science in Context*, 21(3): 377-402.

Weimer, Wolfram. 1995. Wo lebte der erste Europäer? *Frankfurter Allgemeine Zeitung*, September 7: 12.

Weingart, Peter. 1998. Science and the media. *Research Policy* 27: 869-879.

Whitley, Richard. 1985. Knowledge producers and knowledge acquirers. In *Expository science: forms and functions of popularisation*, eds. Terry Shinn and Richard Whitley, 3-27. Dordrecht: Reidel.

CHAPTER EIGHT

MAD DOCTORS, BAD ACADEMICS AND KNOWLEDGEABLE LOCALS: SOME PRELIMINARY OBSERVATIONS ON THE USE OF MEMBERSHIP CATEGORISATION ANALYSIS FOR THE STUDY OF PUBLIC MEANINGS OF SCIENCE

SIMON LOCKE

This paper presents some preliminary observations regarding the use of membership categorisation analysis (MCA) with respect to the public meaning of science (PMS). To begin, MCA is introduced and an argument put forward that the use of membership categories (MCs) involves enthymematic reasoning, a form of reasoning that employs incomplete syllogisms. This is briefly illustrated with the example of the term "public understanding of science" as associated especially with the deficit model. Two major implications are drawn out regarding the use of the MC "scientist" in ordinary, everyday, practical sociological reasoning: that this involves moral judgements; and that it is inventive regarding the predicates or attributes "bound" to the category. These implications are illustrated using two examples, one concerning a characterisation of academics as "bad" because of their claimed breach of a set of normative predicates, and a second involving the characterisation of doctors as "mad" because of a specific moral predicate bound to the "scientist". Finally, a third illustration is used, that of "knowledgeable locals", to suggest that this is also a membership categorisation device (MCD) of ordinary reasoning and that as such its status as a description of the public is no less rhetorical than the terms of the deficit model.

Membership Categorisation Analysis and/as Enthymematic Reasoning

MCA was originally instigated by Harvey Sacks (1995) as the study of the way people use categorisations of themselves and others as part of their ordinary understanding of the social world. Sacks (1995, 113-125) viewed categories as "inference-making machines" used to deduce things about the actions of individuals to whom a category is applied and to apply categories to explain observed actions. His often cited illustration (Sacks 1974; see also Baker 2000; Lepper 2000; Silverman 1998) is the opening line of a story told by a child: "The baby cried. The mommy picked it up". Sacks argued that our ordinary inference is that the "mommy" is the mother of the baby, partly because picking up babies is what mothers do and because a mother will ordinarily do so when her baby cries. As he (1995, 241) put it there are "category-bound activities" (CBAs), actions ordinarily linked to categories that enable such inference-making. Accordingly, as Miller and McHoul (1998, 124) argue, "inferences *go backwards* from categorial explanations to describe events". The mother's action is explained by virtue of the category used to describe her as such. No assumptions need be made about this particular mother's internal mental state, her motivations, intentions, attitudes or personality, as the category alone is sufficient to explain the action rather than anything about the particular person within. The same applies to the baby crying: crying is what babies do, unlike other categories of actor such as, say, academics. This is how MCs are employed ordinarily as "machines" to make inferences about actions, without need to refer to the "ghost in the machine", as it were.

It can readily be seen how this might be applied to a term like "public understanding of science", a phrase containing a membership category, "public" and a CBA, "science", with an implicit contrast between them (see also Cooter and Pumfrey 1994, 253-54). Employing the "inference-making machine", we might infer that "science" is an activity associated specifically with scientists: science is what (only) scientists do. "Public", on the other hand, is a category into which everybody else falls, their membership being defined by the negative quality of not being a scientist. We might readily then be led to infer that the public do not and perhaps cannot understand science. Given this, some version of the deficit model would seem to be an inevitable consequence; arguably, it is built into the terms of the description. It has also been suggested that categories have attached to them further attributes in addition to actions, referred to as "predicates" (Hester 1998, 135), or "category entitlements" (Whalen and

Zimmerman 1990), including ownership rights (Sharrock 1974). Thus, science is not just done by scientists, it is also owned by them. Accordingly, there is a moral order advanced (Jayyusi 1984): the public qua "non-scientists" *should not* claim to be scientists or to do science, because science is an activity and a form of knowledge category-bound to scientists. For members of the public to presume to do science or to assert that their own knowledge is scientific constitutes a potential normative breach (Garfinkel 1967, 35-75) that might provoke strong negative moral sanction, as is clearly apparent in the notion of "anti-science" (Holton 1992) and the overt moral language sometimes employed regarding such groups as creationists (Kitcher 1982; see also Lessl 1988, Cremo 1998).

On the other hand, it is also apparent from this example that the matter is not quite as straightforward as this inference-making version might seem to suggest. As we know, the deficit model version of the relation between science and the public is widely disputed partly because the strong category contrast is rejected. In terms of MCA, the category predicates are disputed on a variety of grounds, or alternative predicates are advanced. For example, it may be disputed that science belongs solely to scientists since science is often described and legitimised as "public knowledge" (Ziman 1968; see also Shapin and Schaffer 1985); or it might be argued that the public also includes scientists at least when outside their home discipline (Lévy-LeBlonde 1992). What this points to is that the inferential reasoning involved in the use of MCs is rather more flexible than Sacks's mechanistic metaphor suggests; specifically, it is enthymematic.

For Aristotle (1946), the enthymeme or incomplete syllogism is the characteristic of rhetorical reasoning that differentiates it from logic: where logic presents a statement of a particular ("Socrates is a man") and a generalised justification ("All men are mortal") followed by a deduced conclusion ("Therefore, Socrates is mortal"), the enthymeme provides only "a statement together with a justification" (Billig 1996, 132) or leaves out premises (Gross 1996, 12), obliging the audience to fill the gaps in reasoning to make its own conclusions. More loosely, enthymematic reasoning is probabilistic, extending an open palm to the closed fist of the ostensibly unarguable syllogism. Thus, the audience's inferred conclusion may have been what was implied, but the rhetor might dispute it, just as the audience might dispute that the justification warrants the inference. In the delightful term Billig (1996) adopts from the 16th century rhetorician Ralph Lever, enthymematic reasoning both exemplifies and invites "witcraft", the inventive disputation of inferences and premises.

MCDs are enthymemes bearing these rhetorical features. Sacks (1972, 32) defines a MCD as

> that collection of membership categories, containing at least a category, that may be applied to some population, containing at least a Member, so as to provide, by the use of some rules of application, for the pairing of at least a population Member and a categorisation device member. A device is then a collection plus rules of application.

This raises the question of how we decide what "collection" a category comes from, a potentially troublesome issue since the same MC may belong to several different collections: "baby", for example, may belong to the collection "family" or "age"–or even, as Silverman (1998, 80) tentatively suggests, "romance". Similarly, as suggested above, the MC "scientist" may belong to the relatively narrow collections "occupation" or "profession", or to the much broader collection "member of the public". MCs, then, are potentially ambiguous. Sacks attempted to specify a set of "rules" and "maxims" by which members resolve such ambiguities in practice, but the extent to which this is fully possible remains disputed (Schegloff 1995, xlii). An alternative view argues that categories should always be treated contextually, such that instead of looking for decontextualised general rules of use, analysis should attend to the specifics of how categories are used on different occasions (Hester and Eglin 1997). Or, to turn this around, members themselves may use ostensibly the same category in different ways on different occasions (or even on the same occasion) implicating different "collections" with different associated predicates. In addition, specific usages may be left ambiguous, whether intentionally or otherwise, or may actively exploit ambiguity for specific purposes.

Sacks himself provides an example of this. As Silverman (1998, 75) points out, the inferences drawn about actions or other predicates from categories may potentially lead to accusations of "prejudice", so one way to avoid this is "to use a category and let others construct the explanation". He cites Sacks's example of the government of the Soviet Union once publishing a list of names of "profiteers" with no further comment–except that all the names were Jewish. Thus, the inference left for the audience of ordinary "comrades" to draw was that the named individuals were profiteers because they were Jews. Stated enthymematically, the list is a series of particular statements of implicit form, "X is a Jew" with an implied generalisation inferentially bound to this that "Jews are profiteers" and the stated conclusion, "X is a profiteer". Neither the initial statement nor the justification were stated explicitly, but they did not have to be because inferential reasoning could be relied on to work back from the conclusion using what "everyone knows" to be true about the category "Jew". But because it is unstated, the inference is also defeasible.

Another example from Sacks (1995, 113) concerning a call to a social service agency shows how this works in a more clearly interactional situation (A is an agency worker, B the caller):

> A: Yeah, then what happened?
> B: Okay, in the meantime she [wife of B] says, 'Don't ask the child nothing.' Well, she stepped between me and the child, and I got up to walk out the door. When she stepped between me and the child, I went to move her out of the way. And then about that time her sister had called the police. I don't know how she...what she...
> A: Didn't you smack her one?
> B: No.
> A: You're not telling me the story, Mr. B.
> B: Well, you see when you say smack you mean hit.
> A: Yeah, you shoved her. Is that it?
> B: Yeah, I shoved her.

Sacks draws attention to A's question, "Didn't you smack her one?" and further insistence that B is not telling the full "story", wondering how A, in the absence of any other information could know this with such certainty that B eventually admits to some physical violence. Sacks argues that A inferred an act of violence occurred because the police were called and a CBA of the police is dealing with violent incidents. Sacks (1995, 116) spells out A's reasoning more fully:

> What we have is roughly something like this: A knows that the scene is 'a family problem.' So (a) is the family quarrel, (b) is the guy moving to the door...(d) is the police coming. And (c) is the grounds for the police to have come. That is, apparently on some piece of information the police have come, and that piece of information is the thing that A has guessed at. A apparently knows, then, what good grounds are for the police to be called to a scene. And he's able to use those good grounds, first to make a guess, and then to assess the correctness of the answer to that guess.

As should be clear, A's reasoning rests on an enthymeme: given the initial premise of a family quarrel and the conclusion that the police were called, A fills in the missing step based on a general justification, the CBA that the police deal with violent incidents. Also apparent is the merely probabilistic character of enthymematic reasoning, as A's first "guess"–that is, a probable supposition or speculation–is denied by B, who only accepts the validity of A's inference after "smack", reformulated as "hit", has been somewhat mollified to "shove". Thus, although A is correct in making the general inference that something happened for the police to be called, the precise grounds are only probable and therefore open to dispute.

It follows that like enthymemes, MCDs lack definitive closure; thus, they are ambiguous and merely probabilistic, and hence always potentially disputable. MCDs, as Silverman (1998, 89) states, are used by people to make "*presumptively* correct descriptions"; "presumptive" because the description can be wrong and what people categorise using one collection might turn out to involve a different collection entirely. Thus, choice of category is significant and given there is always at least one alternative (Sacks 1972), categories do persuasive work by representing reality one way rather than another. Similarly, use of categories is "recipient designed" (Silverman 1998, 89), chosen with a specific audience in mind to persuade them–and maybe also ourselves–to see things in one way rather than another.

All this dovetails neatly with the probabilistic nature of enthymematic reasoning, as also does a further feature Sacks (1995, 42) mentions that the inferential nature of MCDs involves "making new knowledge". In linking predicates to categories and categories to predicates, people infer things about the ongoing events of the social world around them. But the lack of closure means there is always room for uncertainty and the potential that category ascription will be disputed. Thus, people have to be inventive in making category descriptions, both in the initial act of describing and to counter potential questioning of the categorisation advanced. Categorisation, then, is active work that may involve providing justification through elaborating both the description and the meaning of the category itself. Categories do not merely provide resources of narration as in the child's story, but are themselves developed and elaborated narratively.

In addition, a further significant feature of categories is that they implicate a moral order as follows from certain of their predicates. Thus, as Jayyusi (1984) argued, categories are not just descriptive, but also evaluative and judgemental. Again, choice of category is critical in this: to describe someone as a "Hell's Angel", for example, is to position them in relation to the prevailing categorical order and impute a set of behaviours already cast in a moral light, as "good" or "bad", "praiseworthy" or "blameworthy" and so on. None of this is definitive in its specifics as it depends on who is speaking to whom about what, but that there are judgements of value involved follows from the fact that an alternative category can always be used, such as one that might be neutral with respect to a collection like "biker-gangs" or "youth subcultures", but impute instead a different set of moral predicates (for example, "mate", "son", "mechanic", "lover", etc.) (see also Widdicombe and Wooffitt 1995; Widdicombe 1998).

Such moral and other ambiguities are a feature of the category "scientist", which, as in the contrasting representations provided by positivism and Romanticism (Locke 2005, 2011), incorporates moral judgements about whether it is a good or bad thing to be and about good and bad ways of doing science. In addition, the discourse of science itself enables it to be treated as both category-bound and category-free. It can be represented as both particularistic and universalistic (Winiecki 2008); both "science-in-particular" and "science-in-general" (Michael 1992); both modern-western and transcendent of time and space; both "privately" owned by the scientific priesthood and publicly shared by the synecdochic priestly voice (Lessl 1989; Locke 2005, 2011); both alien to and aligned with common sense (Gregory and Miller 1998). MCA can help to unpack this further through its focus on the active work of moral reasoning undertaken by people in their ordinary, common sense-making. In this, the category "scientist" is positioned within a pre-existing general moral order, but not in any simple, uniform or unproblematic way. Rather it is complex, mixed and puzzled activity as ordinary people try to figure out what "scientist" means by defining, clarifying, elaborating and applying its predicates. This creative work is at once both descriptive and moral, involving judgements of "badness" and "madness" that are wrapped up together in narrative bundles that people unpack through enthymematic inferential reasoning.

However, it is not just a case of science and the activities of scientists being positioned in relation to an existing moral order; rather, science itself is used as a resource of moral assessment (Cameron and Edge 1979). This does not mean just that science is interpreted through whatever existing moral framework(s) inform public reasoning, but that the resources provisioned by science are treated as inherently moral in that they are used as bases on which to type people and judge action within an order of social evaluation, one that includes scientists themselves. Thus, ordinary people are actively involved in attempting to work out both where the scientist fits in relation to the existing framework of moral assessment, but also reworking this framework by inventively developing the resources science provides to construct new bases of moral judgement. Thus, the moral ordering informing MCs should not be thought of as a pre-given, pre-defined framework that is simply imposed on people or situations (Hester and Eglin 1997), but rather as an available resource that people use to work out possible moral assessments through the activity of categorising. In rhetorical terms, this involves invention, the creation of arguments using available rhetorical resources–the means of persuasion–to think with, about, and through the meaning of situations in their specificity and immediacy. This reasoning has to be *done*, and done *in situ* as an ongoing practical accomplishment. However, in keeping with the incomplete logic of the enthymeme, the form this reasoning takes is narrative, involving the construction of accounts serving to function as explanations. Thus, people construct speculative stories (Eglin and Hester 1999), the sense of which is assessed by narrative criteria including coherence and credibility (Fisher 1994). Thus, for any given action or observable event to be attributed to a specific category of member, it must be understood as credible that that category might undertake such an action or be responsible for the event happening–unless it is the very *in*credibility of such a category taking such action that is the story. In either case, the action must (be made to) fit the perceived character-type or *persona* (Campbell 1975) and thus make for a coherent story.

To provide some initial illustration of these features, I now turn to two specific usages, one of the category "doctor" represented as a man of science, but first an example of the closely related category "academic", which I shall argue is viewed in context through a set of predicates commonly associated with the "scientist". (Further information about the selection of cases is in Locke 2011.)

Bad Academics

The first example is taken from an article published in the high-street magazine *Fortean Times* (FT). FT is the principal outlet for British followers of the early twentieth century American "philosopher" (their category), Charles Fort. Fort had no academic affiliation, but spent much of his life trawling through newspapers, magazines and scientific journals gathering what he called "the damned", that is "data" that do not fit established (especially scientific) views of reality, summaries of which were published in four books (Fort 1995, 1996, 1997, 1998). Fort's contemporary followers continue this work and the content of FT mainly consists of reports of such "strange phenomena", as its sub-title states and articles discussing various aspects of them. One such article was devoted to reviewing a number of recent books by "academics" on the topic of unidentified flying objects (UFOs) (Brookesmith 2008) and provides a good example of how the moral predicates attached to this MC are mobilised as a resource of public criticism.

As follows from the nature of Fort's project, a general feature of the discourse of Forteans is that it articulates a mixed relationship to conventional science. On the one hand, Forteans are dependent on science as one source of the "strange phenomena" they document; but on the other, they are often "sceptical" (their term) towards conventional scientific understandings and explanations, especially where these are seen as "debunking" Fortean interests (sometimes referred to as "skepticism with a 'k'," in specific reference to the Committee for Skeptical Inquiry, see http://www.csicop.org/about/about_csi/). This mixed relationship is apparent in the way Forteans use institutionalised signifiers of academic expertise to display acceptance and legitimation, while also distancing themselves from such institutional participation or involvement. Thus, Forteans often refer to their activities as "research" and display their conventional scientific qualifications (such as the ufologist, Dr David Clarke) or institutional status (such as Paul Devereux described as "a founding co-editor of the peer-review *Time and Mind: The Journal of Archaeology, Consciousness and Culture*", FT264, July 2010: 20). But they also distance themselves from institutional academic activity and do so using predicates attached to this MC.

This can be seen in Brookesmith's article. To begin with, the article, entitled, "When aliens go to college", is prefigured by an introductory sub-heading stating

> What happens when fortean subjects gets [sic] the academic treatment? Peter Brookesmith looks on in horror as Grays, reptilians and worse invade the groves of academe–and discovers that 'legitimate' scholarship has singularly failed to pay its dues to much of the ufological research that went before. (Brookesmith 2008, 48)

Here, the category "academic" is immediately set up contrastively to "fortean subjects". Thus, a category entitlement in the form of a claim of ownership is advanced over certain "subjects" bound to the category "fortean". However, in common parlance, the collection "subjects" is also bound to academic activity as a description both of general topics and disciplines of study, and specific targets of research. Thus, the ambiguity in the collection is used to invest the category "fortean" with an equivalent status to "academic" as ones to which disciplined research activity is bound, but the contrast between "fortean subjects" and "academic treatment" positions these "subjects" as belonging to Forteans, while only "treatable" by academics. So, whilst academics might deal with such "subjects", they cannot claim ownership over them because they belong to Forteans. From the outset, then, academics are already bounded off from these subjects.

So what are these subjects? Given the title of the article, they are specified in this case by the category "aliens", which within Fortean discourse implies extra-terrestrial beings, often associated with UFOs. The title further cements the above contrast in describing aliens as "going to college", as this location category (Silverman 1998, 134) is made into something aliens *go* to rather than *belong* to. The editorial sub-heading accentuates this through giving further specification to aliens as a collection that includes "Grays, reptilians and worse" (Grays and reptilians being purported types of extra-terrestrial), which have attached to them the CBA "invading" directed at "the groves of academe", a location formulation that gives specificity to the category "college" as a tree-lined idyll. Further, the "invading" is given explicit negative evaluation through the three-part list (Wooffitt 1992) of types of aliens ending in "worse" and so implying a succession of increasing monstrousness, specified as "horrors" witnessed by Brookesmith. Thus, the sub-heading plays on a double-meaning of "horror" as a general predicate of "invading aliens", to which Brookesmith might then have the expected reaction of any ordinary human being, but also as his reaction to "fortean subjects" being "treated" by academics. Is his horror then in sympathy with the academics or the fortean subjects? The answer becomes immediately clear as it is made explicit that "'legitimate' scholarship" has "failed...ufological research". Thus, the contrast structure is given closer specification still through this opposition, which again mobilises category entitlements to attach ownership of "research" to "ufology" positioned against the "legitimacy" attached to "scholarship". However, the scare quotes around "legitimate" cast this scholarship in a dubious light so that it appears rather more *il*legitimate and, from the sense given of what it has failed to do–"pay its due"–specifically as stealing.

This stealing follows from the attachment of the subject of aliens to the category "fortean" without requiring any further specification of the activity of academics. However, the article goes on to provide such specification by drawing on certain CBAs attached to "academic", which are used to cast the activity of some particular academics in a negative light. These can be brought out from the criticisms made in the article. Listed in order of appearance, these include the following (numbers in brackets refer to the pages of the article):

1. when academics of an anthropological or sociological disposition start to take an interest in a cultural phenomenon, one can take a reasonably safe bet that, as a cultural entity, it's all but dead (49)
2. not entirely new thinking (50)
3. One began to wonder...how much and how closely academics really did read and, worse, whether they were simply careful about what they *admitted* having read, lest they appear insufficiently original (50)
4. ignorant of history (51)
5. unaware of long-standing pragmatic resolutions of the theoretical difficulties of democracy (51)
6. lack of both compass and moorings...leads to utter 'relativist' fatuities (51)
7. an opaque postmodern version of what [a ufologist], [a] fortean...and at least one leftist cultural critic...had said in plainer English before (51)
8. saying nothing that veteran readers of *Fortean Times* and more specialist publications hadn't heard a score or more times before (51)
9. she's an academic: she's paid to read (51)
10. when scholars without official academic standing were finding it all but impossible to get their thoughts on ufology turned into books, three writers from various levels of the ivory tower garnered attention (51)
11. decided, arbitrarily it seemed, that memories of abduction were almost certainly false (52)
12. makes zero effort to defend her judgement (52)
13. without tying them to specific cases in the literature (52)
14. Incredibly, [a ufologist's] work on this question is mentioned nowhere in her book. Yet she claims to have read 'every account of alien abduction ever published'. This is not apparent from her embarrassingly inaccurate summaries of the few particular cases she does address. (52)
15. in general fails to follow them through with analyses and conclusions from salient cases known to all and sundry. Worse, she seems to be oblivious of the work of a whole collection of less than dewy-eyed ufologists (52)
16. [Her] ufological references tend to be bipolar: to true believers or to extreme sceptics. This particularly won't do, because most of [her] 'discoveries' about the nature of abductions...have been aired by [ufologists] (52)
17. Disappointingly, but unsurprisingly, [she], who has 'read everything', doesn't cite the book in notes or mention [it] in her index (53)
18. read widely among the kind of scholarly ufologists named.... But not, apparently, comprehensively (53)
19. the books and articles...that directly anticipate her conclusions, go unmentioned, while her notes would lead one to think that what she has read of them she has deemed irrelevant (53)
20. one would like to see credit given where it is due (53)
21. playing fast and loose with sources (53)

22. these academic books...tend to treat the abduction (and UFO) phenomenon as almost exclusive to the continental United States (53)
23. ufological scholars will lead the way and break the ground, and academics will quietly follow after, and build reputations on the back of unsung pioneers (53)

The notable feature of these criticisms is that they draw heavily on a set of CBAs associated with scientists that are broadly similar to Merton's (1968) ideal-typified description of the norms taken to characterise the scientific community as: universalistic; communistic; disinterested; and sceptically organised. Basically, these mean that scientists should assess knowledge claims not claimants, share knowledge communally, strive for objectivity, and avoid reaching conclusions until all the evidence has been reviewed. Brookesmith finds fault with academics on each of these grounds as follows (numbers refer to the above criticisms):

Universalism. Instead of being universalistic, they are particularistic in their sourcing of ideas, specifically excluding or failing to credit "ufologists" and/or "forteans" (7, 8, 14, 15, 16, 17, 19, 20).

Communism. Instead of treating knowledge as a public good, they are socially isolated and exclusive (1, 7, 8, 9, 10).

Disinterestedness. Rather than being disinterested, they are motivated by personal interests focused on priority and reputation (3, 23).

Organised scepticism. Far from being sceptical, they are ignorant of the evidence and dogmatic in their views (2, 3, 4, 5, 6, 11, 12, 13, 14, 15, 16, 17, 18, 19, 21, 22).

Thus, akin to other critics of conventional science such as creationists (Prelli 1989; see also Locke 1999), Brookesmith mobilises CBAs associated with the category "scientist" to find the work of academics wanting and to view them as morally culpable in comparison to ufologists and Forteans, who by contrast appear as proper "scholars" nobly pursuing ground-breaking research with no concern for institutional reputation or standing. In being represented as breaching the norms associated with their categorisation, these academics then appear bad, failing to live up to their presumed categorical billing. In Mulkay's (1979, 71) terms, the CBAs of "academic" are drawn on as "standardised verbal formulations" providing rhetorical resources that, while they may be used to defend scientists' actions, can also be wittily turned against particular academics as means of moral assessment. Viewed enthymematically, the unstated syllogistic logic at work is: "Academics are universalistic, communistic, disinterested, and sceptical"; "K is an academic"; "therefore, K is universalistic, communistic, disinterested, and sceptical". Thus, making the case that K is not one or other of these is a way of finding fault with K and undermining the academic credibility of their work. The particular individual is assessed by how they meet up to the category in terms of the presumptive predicates attached to it; it is the category that is taken to be the measure of action and the activities of individuals are "treated" (pun intended) in these terms. The category provides rhetorical resources for moral assessment and the activities of individuals are both described and judged accordingly. Further, this does not merely apply to measures of "badness", but also to "madness" as the next example illustrates.

Mad Doctors

The above example shows that MCs provide a basis of normative expectation and assessment, a moral base-line against which individual action can be judged. The bases of judgement are not just moral, but also rational. MCs define what sort of behaviour is reasonably to be expected from members to whom a category is applied; or to turn this around, where activities occur that are deemed to be beyond the norm, some category of member is looked for that could potentially or conceivably perform such actions by virtue of the predicates normatively bound to it. As Sacks put it (1995, 42), people are continuously monitoring events "to find out what is getting done by members of…categories". In consequence,

> we get nicely special kinds of occurrences which provide a beautiful view of tensions arising as persons await the discovery of which of them is going to be found to have done this thing.

In the case of actions viewed as beyond the norm, the work involved for people in the first instance is to identify potential categories by matching features of the events, itself involving descriptive narration, to known predicates of categories to try to figure out what was done, how it was done and by whom in the sense of what sort of category. Identifying categories then enables and organises search procedures to find the particular individual(s) responsible, as well as disputes over whether or not a given individual could or could not be responsible (Sacks 1995, 180).

An example that illustrates this is the case of Jack the Ripper in which amongst the categories identified as potential suspects was–and still is–

"doctor". The Ripper case is particularly interesting as it "offers no closure" (Walkowitz 1992, 201), because the crimes associated with the Ripper have never been solved. As Frayling (1986, 214) put it, Jack the Ripper is an "absence", a name we all know for someone nobody knows; indeed, there is a sense in which the name refers less to a particular person and more to a set of events, specifically the murders of a number of women in the district of Whitechapel in the East End of London in the Autumn of 1888. There is no space here, nor any great need, to go into full detail about the murders, except to say that they showed what is often said to be a common modus operandi, including the type of wounds and hence the sort of weapon employed–a knife long and sharp enough to eviscerate and almost decapitate. As these details became publicised, speculation about the possible culprit(s) turned from what might be called "the usual suspects"–that is MCs typically associated with criminal activity, such as local street gangs (although suspicion also focused on Jews)–to other kinds of category not typically associated with murder. Thus, as the descriptions of the murders became more extreme, drawing on what Walkowitz (1992, 196) calls the resources of the "fantastic", so speculation about the possible MC responsible also reached toward the fantastically monstrous: as the going got monstrous so the monstrous was got going.

Significantly, prominent amongst such fantastic figures was the category "doctor" represented as a man of science. The MC "doctor", then, specifically as a type of scientist has some predicates bound to it that enable it to provide for people a recognisably possible category to fill the absence that is Jack the Ripper, to be the category that did this thing. Something of what these predicates are can be discerned from press reports of the time, which I have drawn from the detailed account provided by Walkowitz (1992). What follows then should not be seen as a full MCA of "Jack the Ripper", but rather an initial attempt to explore some features of the inferential reasoning involving the categories "doctor" and "scientist" with a view to considering what this suggests about PMS. Specifically, it suggests that doctors are viewed as marked by a moral character that is uncertain and compromised because of their category-bound association with an instrumental mode of action, which is viewed as both "bad" but more so as "mad". In addition, the resources of science may be used as a basis of sense-making, from which categories are derived that are given moral weighting and used with ironic wit to tell possible "atrocity stories" (Dingwall 1976) about doctors as men of science.

As already suggested, as the murders that came to be associated with the Ripper continued, press coverage moved from a focus on "the usual suspects" local to Whitechapel to take in a wider range of possible suspects, including members of what Walkowitz calls "respectable" society. After details of the murder of Annie Chapman in particular were made public, Walkowitz (1992, 205) describes the press reportage as more "sensational" making references to "'bloodthirsty' monsters and fiends", that is MCs of an extreme and unusual nature. This suggests that the boundaries of "normal deviance", so to speak, had been breached and that the lack of closure enabled and encouraged speculation to stretch beyond established limits of the known to invent new grounds of the knowable. Significantly, the category "doctor" figured prominently in this speculative territory. It is notable that this was fuelled by a member of the medical profession, the police surgeon at Chapman's inquest, who stated that

> the work was that of an expert–or one, at least, who had such knowledge of anatomical or pathological examinations as to be enabled to secure the pelvic organs with one sweep of the knife. (Quoted Rumbelow 2004, 42)

Such a judgement coming from a doctor presented the possibility of inferential reasoning along the lines of "takes one to know one". That a skilled practitioner was declaring the activity to be recognisably skilled opened the possibility that the murderer was from the same category. The work of creating new knowledge then focused on generating speculative narratives in which a doctor could play the role of villain.

One source of speculation with notable significance for PMS came from Stevenson's proto-science fictional story, *Doctor Jekyll and Mr. Hyde*, a stage version of which happened to be running in London at the time of the murders. This presented Hyde as sexually motivated, a monstrous and violent "manifestation of Jekyll's lust" (Walkowitz 1992, 206). The suggestion was then made that the Ripper might be "an evolutionary throwback" or "troglodyte" driven to violent murder by "erotomania", a notion recently popularised from Krafft-Ebing's casebook of sex offenders, *Psychopathia Sexualis*. Such a blurring of the boundary between "fact" and "fiction" suggests that the Ripper's crimes were perceived as having torn the fabric of the normative order to such an extent that the boundary of the "real" became uncertain and what might have seemed impossible began to seem, not merely possible, but perhaps even probable. For some at least, the fictional imaginings of the possible activities of scientists began to seem quite reasonable inferences. Thus, the Ripper was also cast as a "crazed biologist" or "mad physiologist" (Walkowitz 1992, 207), with a further notion of "syphilitic madness" and the speculation that the Ripper was a "medical student" avenging himself on those from whom he had contracted the disease (Walkowitz 1992, 210).

Sexual motivation, however, was only one source of speculation and the papers

> debated whether the murderer…was mad or vicious, a victim of disease or a practitioner of 'mere debauchery', a 'homicidal maniac' bent on violence or an 'erotomaniac' bent on sexual satisfaction. (Walkowitz 1992, 208)

Further proposed categories included "human vivisectionist" and "bodysnatcher", while another type of medical motivation, suggested by the coroner at Chapman's inquest and described by him as more "rational", was that the Ripper was harvesting a uterus to sell as there was "a market for the missing organ" (quoted Walkowitz 1992, 209). There were also other types of "privileged villain" proposed, including one further type of scientist, "the Scientific Sociologist" (Walkowitz 1992, 212).

From Walkowitz's account of the press reportage, then, a range of types of "mad doctor" were advanced as putative identities for the Ripper, including: "evolutionary throwback/troglodyte"; "homicidal maniac"; "erotomaniac"; "vengeful syphilitic"; "body-snatcher"; "vivisectionist"; and "scientific sociologist". How do these categories work as projected resolutions of the tension generated over who it was that did this thing? As discussed above, to provide a means of potential closure, the category needs to be linked to biography and the suspicion that a doctor might have been involved raises an immediate problem, since murder is not typically a CBA of this occupational category. Thus, inferential speculation was led to consider further predicates possibly associated with "doctor" that might provide a credible and coherent narrative to fill the missing step(s). What is notable about the prospective answers advanced is that they drew from recognisably known things about scientists in two different ways: one used the stories scientists themselves were beginning to construct about the nature of humanity in general to apply these with ironic wit to the category "scientist"; the other drew on predicates specifically bound to this category with one common characteristic–instrumentalism.

In the first case, then, we have "evolutionary throwback". This explicitly draws on the scientific view that human beings have evolved from "primitive" species of hominid, which then enables the speculative notion that prehistoric "troglodytes" might somehow reappear as a "throwback" of reproduction or as a result of some bizarre scientific experiment. Given that the populace of Whitechapel were themselves likened to "savages" by "respectable" commentators (Rumbelow 2004, 5; see also Curtis 2001; Walkowitz 1992, 197), it became possible to imagine such a "primitive" appearing even amongst "cultivated intellect[s]"

(Walkowitz 1992, 208). Thus, I would suggest, this category draws from the collection "race" made available by the discourse of science, a collection that defines "otherness" on the basis of species-membership and implicates a moral order in which species are ranked in terms of their presumed level of development.

Meanwhile, the categories "homicidal maniac", "erotomaniac" and "vengeful syphilitic" use a different means of defining "otherness", but also draw on an emerging scientific discourse: psychopathology, which defined madness in medical terms and pursued systematic classification of its forms (Foucault 2001). Thus, just as the notion of "erotomania" was taken from Krafft-Ebbing, so was the idea of a specific connection between venereal disease and madness taken from then current medical science (Walkowitz 1992, 210). These categories, then, draw from a different collection, "mental illness", but one again made available by the discourse of science. This implicates a moral order directed at an individual level rather than a whole species, providing a different basis on which to judge action with different possible consequences and outcomes, as well as alternative narrative scenarios.

However, although they implicate different kinds of moral order, both the collections "race" and "mental illness" partake of science's universalistic rhetoric and as such are ways of categorising everybody, not just scientists. Thus, applying them to doctors shows ordinary people's inventive witcraft, turning the tables on scientists by using their own rhetoric to cast moral judgement presumptively on a suspected member of the scientific community itself. The remaining categories, however–"body-snatcher", "vivisectionist" and "sociologist"–show a further difference, as they refer to activities specific of or directly linked to scientists. "Body-snatchers" procured human corpses to sell to medical researchers and the suggestion was made that the Ripper was doing likewise with female organs; similarly, "vivisectionists" practised their research on live animals and the horror of the Ripper was the possibility that he might be doing the same to women. In either case, activities bound to medical researchers were being drawn on, activities of an essentially instrumental nature in their treatment of human beings. Whether the Ripper was reaping organs for cash or knowledge for medicine, he was treating women as a means to an end. The "scientific sociologist", too, shows a similar morally compromised instrumentalism. The category, proposed by Bernard Shaw "facetiously" according to Walkowitz (1992, 212), was described as "a social reformer trying to expose the conditions of the East End". Facetious or not, it is significant that it *could* be proposed as a CBA of the scientist, treating ordinary people's lives as a means to an end, even if with "good" moral intent. Thus, the collection for these categories is "scientist" and forms of instrumental action are viewed as a CBA of this category. But this is not just a description, it is also a moral judgement in which the scientist appears as an "other", a category outside the existing moral order and as such of uncertain and potentially dangerous status. The scientist's instrumental ethic, therefore, is both category bound and bound to no good.

It must be stressed again that this analysis is provisional and intended to be no more than suggestive of the possibilities of MCA for studying the category "scientist" in PMS (see also Winiecki 2008). In this respect, there are three main points to bring out.

First, it provides some support for the more general argument made elsewhere (Locke 2002) that PMS is rhetorically inventive, particularly involving enthymematic reasoning. The above analysis shows that people can and sometimes do draw on science as a resource, using it to construct prospective descriptions of action that then have a "scientific" character to them, albeit of a speculative and hence arguable form. People do use science to construct understandings of actions and events in the social world; however, these understandings may or may not conform to those currently advocated by members of the scientific community (itself speaking a multiplicitous discourse). The reason for this is not necessarily a matter of "ignorance" or other form of "deficit", but because people use science as one source of materials to fill the gaps to account for observed actions and thereby prospectively seek to resolve the tension in finding out what category did the thing. Accordingly, science is treated as a resource of moral reasoning, a means with which to construct categories of potential actor, who are always already positioned morally by virtue of the action to which they are linked. Thus, the relationship between science and the existing moral order(s) is dynamic and dialogical: science provides new categories to build into the existing moral order which becomes reconfigured accordingly; but the order remains a moral one and the scientific categories are positioned within its system of relative valuations. This dynamic is fuelled by the inventive use of science to tell stories about both the possible actions of known categories and the possible categories behind known actions. This is done with witcraft, using ordinary skills of rhetorical reasoning to make a persuasive case for perceiving the matter one way rather than another. A notable feature of this witcraft in the case of Jack the Ripper was to use the universalistic rhetoric of science against scientists, cleverly arguing that a prospectively culpable man of science was himself suffering the effects of a scientifically conceived degeneracy, whether of a species-wide or individual nature. Whatever such stories may have lacked in scientific credibility they made up for in their ironic narrative coherence.

Second, the "scientist" as a category may itself be represented as a moral "other" by virtue of a form of activity bound to it: instrumental action. Such action is a focus of moral opprobrium through a variety of negatively evaluated category types linked to it, all of which are directly associated with the doings of scientists. This is not to argue that this is always the case and it may be that in some contexts acting instrumentally is given positive moral assessment, as cause for praise rather than blame. Nonetheless, the significant point is that it can be a basis of negative judgement and that it is morally judged at all. In being conceived as amoral, instrumental action is thereby judged morally. Thus, if this is treated as a predicate of the category "scientist", then this category is assessed in relation to existing moral standards. Accordingly, the relationship between science and morality is dynamic and scientists may be held accountable by and to a moral order that, while lacking formal codification, is that much more powerful by virtue of its informality and everyday constitution. In relation to this, the scientist stands as an uncertain, ambivalent and somewhat mysterious figure, both somehow known and not known. The scientist, then, as an occupational category or profession is "other", bearing predicates of a potentially monstrous form. It is important to recognise that this is about *categories*, not biographies; the categorical order has its own workings regardless of the individuals that fill category positions. Exceptions can always be made without calling the categorical order into question (Jayyusi 1984; see also Billig 1996, 148-185); rather, the categorical order is used to judge individuals as "deviant", whether bad or mad. Occupations are part of a categorical order to which all adult members belong (even if only by virtue of being "unemployed", "retired", "housewife"/"househusband", etc.) and within which every occupation is a potential "other" to *any* other (Parker 1995)–so journalists, for example, might readily consider scientist-doctors quite other to themselves (see van Dijck 1998). Each occupation has a different set of predicates attached, enabling different inferences to be drawn and thus potentially different stories to be told about each. Thus, the categories at work need to be studied in their specifics to understand what predicates are employed and in what ways, and how then the "doctor" and/or "scientist" is morally positioned and re-positioned on each occasion–"each next first time" (Garfinkel 2002).

Given this, it might then be argued that there is little to be learned about contemporary views of the "scientist" from the Ripper case. However, the view of the culprit as a "mad doctor" continued to be played out in popular culture over the course of the 20th century (Frayling 1986) and remains current in the early 21st, for example, in the comic-book (or graphic novel) *From hell* (Moore and Campbell 2000; see Locke 2009), made into a film in 2001. Moreover, there are grounds for arguing that "mad doctors" have become both more common and more extreme figures in popular depictions of the scientist (Skal 1998). Here, however, the point about specificity does count, as we should be wary of assuming that every such figure is more or less the same (Roach 2011, this volume). So, while the Ripper case provides some support for the points above, these need to be treated as speculative generalisations providing possible bases for further enquiry into the empirical detail of other cases. They encourage an approach to such enquiry using MCA that avoids assuming we already know what predicates the category "scientist" (or "doctor") has bound to it, or that any one use is the same as any other, but instead focuses on specific usages in specific contexts to begin to build up a more informed and detailed picture of mundane moral reasoning about science. Seeing the "scientist" in instrumental terms is only one dimension of this mundane moral reasoning; as such, it is itself a rhetorical construct providing only one enthymematic narration of the public meaning of science. This contributes to PMS, but public and popular inventive wit reaches far beyond.

Knowledgeable Locals

One of the places it reaches to is a notion of knowledgeable locals. The significance of this to PUS is well established in one of the strong arguments against the deficit model that it underestimates the important role played by local knowledge in people's perception of and relationship to institutionalised science, and further that it misrepresents the nature of such knowledge as "deficient" (Wynne 1991, 1992; see also Irwin and Wynne 1996; Ziman 1991). Not only has this been important in countering the over-simplications of the deficit model, but also the view that the public's relation to science should be understood principally in terms of "expertise" (Collins and Evans 2002; Wynne 2003). What is apparent from these debates, however, is that the notion of "local" is ambiguous, referring at times to geographical location (e.g., Cumbrian farmers) and at other times to matters of identity taken to have certain political implications (e.g., Wynne 1993, 2008; see also Irwin 1995). Whilst this lends the category a valuable flexibility as a means of persuasion within these academic debates, it also shows its enthymematic character. Much like the category "scientist", the predicates that may be attached to "local" are mixed and uncertain; in part, it may be treated as a "location formulation" (Silverman 1998, 134) and in part as a category entitlement incorporating a variety of rights, including ownership over some kinds of knowledge especially that related to the locale in question. It is this mix of predicates that enables the academic debates to be had as they raise or challenge missing propositions that are presumptively taken to apply. In using the term "local", both certain kinds of knowledge and certain kinds of entitlement may be presumed but not necessarily stated; hence, arguments can and have occurred over these presumptions.

"Local knowledge", then, is an MCD, that is, a *resource* of ordinary inferential reasoning that some academics have employed, with unintended irony, to *describe* ordinary inferential reasoning as itself a resource of "expert" argument. Some support for this comes from an example of the use of this MCD in, ostensibly, a non-academic context. Moreover, it is used to challenge what is sometimes represented in this kind of context as a more "expert" view–just as it is used in the academic debates about PUS. The example again comes from *Fortean Times*, from an article discussing reports made in late 1980 by US Air Force personnel stationed near Rendlesham Forest in East Anglia, of encounters with UFOs partly in the form of unusual lights in the sky (Randles and Clarke 2005). Such encounter reports are routinely subject to "debunking", that is efforts to find possible mundane objects that can account for the reported experience, so making the object no longer *un*identified but determinately identified (Dean 1998, 34-38; Denzler 2003; Westrum 1977). In the Rendlesham case, one such object was the Orfordness lighthouse, which was made a point of specific focus by one FT writer, who describes how she "went straight to Rendlesham to check out the theory" (Randles 2005, 38). In her ensuing discussion, she (2005, 38-39) points to a "problem" with the lighthouse as a mundane solution

> because the locals all knew it was there and saw it night after night. There were many local witnesses to strange things that weekend–not just nervous airmen in a dark, unfamiliar forest.

Here, then, Randles draws on the MC "locals" as a resource to warrant the claim that there is a "problem" with the theory that the lighthouse could explain the purported UFO encounters. Specifically, she contrasts "local witnesses" with "nervous airmen in a dark unfamiliar forest", thereby investing locals with greater credibility as observers, partly because of their relative knowledge about the lighthouse, not just "knowing it was there" but "seeing it night after night", so implying "familiarity" in contrast to the description of the "airmen". In addition, the locals also appear as more objective, as sure "witnesses" in comparison to the airmen whose "nervousness", exacerbated by a "dark, unfamiliar forest" might have adversely affected their capacity to observe reality correctly. This second contrast is of note in relation to an argument often made regarding UFO reports that those presented by "trained observers", a collection often taken to include such occupational groups as pilots and military personnel, are more reliable than those made by ordinary people (Dean 1998, 44-45). The argument trades on a predicate of such categories that they are more "expert" in the activity of observing. Here, however, Randles reverses this by drawing on a predicate of locals that they are more knowledgeable in the specific sense of being familiar with the local vicinity and therefore better witnesses of "strange things". "Local", then, as a category of ordinary inferential reasoning has attached to it a predicate of "knowledgeability" in respect of the immediately surrounding environment that can be used to present a counter basis of "expertise" even to those who might be taken to be "trained observers". Randles uses the category in this way to sustain the potential reality of the "strange things", suggesting that, by virtue of their familiarity with the locale and specifically the Orfordness lighthouse, locals would have discounted this as an adequate explanation of what they witnessed. Thus the encounter reports made by the airmen are lent credibility, despite their "unfamiliarity" and "nervousness", because they are backed up by surer observers–the locals. In contrast to such knowledgeable locals, any outside observers, including would-be "debunkers", whatever the basis of their own expertise, might be made to appear relatively ignorant–just as might government scientists called in to investigate levels of radiation in a locality such as Cumbria (Wynne 1993). The inferential reasoning involved is the same and it is based on enthymemes enabling a presumptive contrast between the MCs "local" and "expert".

Conclusion

This paper has sought to advance some initial suggestions about the use of MCA in the analysis of PMS. On the basis of an argument that MCs as features of ordinary, everyday reasoning are enthymematic in character, two examples involving the categories "academic" and "doctor" were considered to show how predicates presumptively linked to the category "scientist" provide inferential resources of public criticism to judge the actions of academics as "bad" and those of doctors as "mad". These examples show both that the activity of scientists is assessed on the basis of an existing moral order within which predicates attached to this category are located, notably that of "instrumentalism", and also that science provides new categorical resources that members of the public can employ with mundane witcraft, to invent new bases of moral assessment to apply to scientists themselves. This makes for a dynamic relationship between science and the existing moral order, suggesting that these predicates and categories are continuously under development through the everyday practical work of inferentially applying them, with potential variations from context to context and so demanding detailed study in each case.

In addition, through consideration of a third example, it is argued that the category of "locals" has attached to it the predicate of "knowledgeability" and, accordingly, that the notion of "local knowledge" is a resource of ordinary, common-sensical inferential reasoning. Thus the use of this MCD in critiques of the deficit model of PUS is itself an example of such ordinary reasoning. It uses everday inferential reasoning as a resource to construct a critique in defence of everyday inferential reasoning, i.e., "local knowledge". Thus, it uses as a resource what should be treated as a topic of study–and the danger of this is that we miss what is going on in such everyday inferential reasoning as we attempt to make it fit into our presumptive categorisation of how such inferential reasoning operates and what it ought to be doing.

Acknowledgements

Some of the material in this chapter has been previously published in *Re-crafting rationalization: enchanted science and mundane mysteries* (Farnham: Ashgate, 2011). An earlier version of the paper was given at the Science and the Public Conference, Imperial College, London, 3-4 July, 2010. My thanks to organisers and participants.

References

Aristotle. 1946. *The works of Aristotle translated into English: Volume XI. Rhetorica*. Trans. W.R. Roberts. London: Oxford University Press.

Baker, C. 2000. Locating culture in action: membership categorisation in texts and talk. In *Culture and text: discourse and methodology in social research and cultural studies*, eds. A. Lee and C. Poynton, 99-113. London: Allen and Unwin.

Billig, M. 1996. *Arguing and thinking: a rhetorical approach to social psychology.* Second edition. Cambridge: Cambridge University Press.

Brookesmith, P. 2008. When aliens go to college. *Fortean Times* 238, July: 48-53.

Cameron, I., and D. Edge. 1979. *Scientific images and their social uses: an introduction to the concept of scientism*. London: Butterworths.

Campbell, P.N. 1975. The *personae* of scientific discourse. *Quarterly Journal of Speech* 61: 391-405.

Collins, H.M., and R. Evans. 2002. The Third Wave of science studies. *Social Studies of Science* 32(2), 235-296.

Cooter, R., and S. Pumfrey. 1994. Separate spheres and public places: reflections on the history of science popularisation and science in popular culture. *History of Science* 32: 237-267.

Cremo, M.A. 1998. *Forbidden archeology's impact: how a controversial new book shocked the scientific community and became an underground classic*. Los Angeles: Bhaktivedanta.

Curtis, L.P. 2001. *Jack the Ripper and the London press*. New Haven: Yale University Press.

Dean, J. 1998. *Aliens in America: conspiracy cultures from outerspace to cyberspace*. Ithaca: Cornell University Press.

Denzler, B. 2003. *The lure of the edge: scientific passions, religious beliefs and the pursuit of UFOs*. Berkeley: University of California Press.

Dingwall, R. 1976. Accomplishing profession. *Sociological Review* 24: 331-349.

Eglin, P., S. and Hester. 1999. Moral order and the Montreal massacre: a story of Membership Categorisation Analysis. In *Media studies: ethnomethodological approaches*, ed. P.L. Jalbert, 195-230. Lanham: University Press of America.

Fisher, W.R. 1994. Narrative rationality and the logic of scientific discourse. *Argumentation* 8: 21-32.

Fort, C.H. 1995. *The book of the damned*. Revised edition. London: John Brown.

—. 1996. *New lands*. Revised edition. London: John Brown.

—. 1997. *Lo!* Revised edition. London: John Brown.

—. 1998. *Wild talents*. Revised edition. London: John Brown.

Foucault, M. 2001. *Madness and civilization: a history of insanity in the age of reason*. Trans. R. Howard. London: Routledge.

Frayling, C. 1986. The house that Jack built: some stereotypes of the rapist in the history of popular culture. In *Rape*, eds. S. Tomaselli and R. Porter, 174-215. Oxford: Basil Blackwell.

Garfinkel, H. 1967. *Studies in ethnomethodology*. Englewood Cliffs: Prentice-Hall.

Gregory, J., and S. Miller. 1998. *Science in public: communication, culture and credibility.* New York: Plenum Press.

Gross, A.G. 1996. *The rhetoric of science.* Second edition. Cambridge, MA: Harvard University Press.

Hester, S. 1998. Describing "deviance" in school: recognisably educational psychological problems. In *Identities in talk*, eds. C. Antaki and S. Widdicombe, 133-150. London: Sage.

Hester, S., and P. Eglin. 1997. Membership Categorisation Analysis: an introduction. In *Culture in action: studies in Membership Categorisation Analysis*, eds. S. Hester and P. Eglin, 1-23. Washington DC: International Institute for Ethnomethodology and Conversation Analysis and University Press of Washington.

Holton, G. 1992. How to think about the "anti-science" phenomenon. *Public Understanding of Science* 1: 103-128.

Irwin, A. 1995. *Citizen science: a study of people, expertise and sustainable development*. London: Routledge.

Irwin, A., and B. Wynne, eds. 1996. *Misunderstanding science? The public reconstruction of science and technology.* Cambridge: Cambridge University Press.

Jayyusi, L. 1984. *Categorisation and the moral order*. London: Routledge and Kegan Paul.

Kitcher, P. 1982. *Abusing science: the case against creationism.* Cambridge, MA: MIT Press.

Lepper, G. 2000. *Categories in text and talk: a practical introduction to categorisation analysis*. London: Sage.

Lessl, T.M. 1988. Heresy, orthodoxy, and the politics of science. *Quarterly Journal of Speech* 74: 18-34.

—. 1989. The priestly voice. *Quarterly Journal of Speech* 75: 183-197.

Lévy-LeBlond, J.-M. 1992. About misunderstandings about misunderstandings. *Public Understanding of Science* 1(1): 17-21.

Locke, S. 1999. *Constructing "The Beginning": discourses of creation science.* Mahwah, NJ: Lawrence Erlbaum Associates.

—. 2002. The public understanding of science–a rhetorical invention. *Science, Technology, & Human Values* 27(1): 87-111.

—. 2005. Fantastically reasonable: ambivalence in the representation of science and technology in super-hero comics. *Public Understanding of Science* 14(1): 25-46.

—. 2009. Considering comics as medium, art and culture–the case of *From Hell. SCAN–The Journal of Media, Art and Culture* 6(1). http://scan.net.au/scan/journal/display.php?journal_id=127.

—. 2011. *Re-crafting rationalisation: enchanted science and mundane mysteries*. Farnham: Ashgate.

Merton, R.K. 1968. Science and democratic social structure. In *Social theory and social structure.* Second edition, 604-615. New York: Free Press.

Michael, M. 1992. Lay discourses of science: science-in-general, science-in-particular, and self. *Science, Technology, & Human Values* 17(3): 313-333.

Miller, T., and A. McHoul. 1998. *Popular culture and everyday life*. London: Sage.

Moore, A., and E. Campbell. 2000. *From hell.* London: Knockabout.

Mulkay, M. 1979. *Science and the sociology of knowledge*. London: Unwin Hyman.

Parker, M. 1995. Working together, working apart: management culture in a manufacturing firm. *Sociological Review* 43: 518-547.

Prelli, L.J. 1989. The rhetorical construction of scientific ethos. In *Rhetoric in the human sciences*, ed. H.W. Simons, 48-68. London: Sage.

Randles, J. 2005. Rendlesham: evolving theories. *Fortean Times* 204, December: 38-39.

Randles, J., and D. Clarke. 2005. Down to earth…. *Fortean Times* 204, December: 32-39.

Roach, K. 2011. Between magic and reason: science in 19th century popular fiction. PhD diss., University of Nottingham.

Rumbelow, D. 2004. *The complete Jack the Ripper: fully revised and updated*. Harmondsworth: Penguin.

Sacks, H. 1972. An initial investigation of the usability of conversational data for doing sociology. In *Studies in social interaction*, ed. D. Sudnow, 31-74. New York: The Free Press.

—. 1974. On the analysability of stories by children. In *Ethnomethodology: selected readings*, ed. R. Turner, 216-232. Harmondsworth: Penguin.

—. 1995. *Lectures on conversation volume 1*. Oxford: Blackwell.

Schegloff, E. 1995. Introduction to *Lectures on conversation volume 1*, by H. Sacks, ix-lxii. Oxford: Blackwell.

Shapin, S., and S. Schaffer. 1985. *Leviathan and the air-pump: Hobbes, Boyle, and the experimental life.* Princeton, NJ: Princeton University Press.

Sharrock, W.W. 1974. On owning knowledge. In *Ethnomethodology: selected readings*, ed. R.Turner, 45-53. Harmondsworth: Penguin.

Silverman, D. 1998. *Harvey Sacks: social science and conversation analysis*. Cambridge: Polity.

Skal, D. 1998. *Screams of reason: mad science and modern culture*. New York: W.W. Norton.

van Dijck, J. 1998. *Imagenation: popular images of genetics*. Basingstoke: Macmillan.

Walkowitz, J. 1992. *City of dreadful delight: narratives of sexual danger in late-Victorian London*. London: Virago.

Westrum, R. 1977. Social intelligence about anomalies: the case of UFOs. *Social Studies of Science* 7: 271-302.

Whalen, M., and D. Zimmerman. 1990. Describing trouble: practical epistemology in citizen calls to the police. *Language in Society* 19: 465-492.

Widdicombe, S. 1998. "But you don't class yourself": the interactional management of category membership and non-membership. In *Identities in talk*, eds. C. Antaki and S. Widdicombe, 52-70. London: Sage.

Widdicombe, S., and R. Wooffitt. 1995. *The language of youth subcultures: social identity in action*. Hemel Hempstead: Harvester Wheatsheaf.

Winiecki, D. 2008. The expert witness and courtroom discourse: applying micro and macro forms of discourse analysis to study process and the "doings of doings" for individuals and society. *Discourse and Society* 19(6): 765-781.

Wooffitt, R. 1992. *Telling tales of the unexpected: the organisation of factual discourse*. Hemel Hempstead: Harvester Wheatsheaf.

Wynne, B. 1991. Knowledges in context. *Science, Technology, & Human Values* 16: 111-121.

—. 1992. Public understanding of science research: new horizons or hall of mirrors? *Public Understanding of Science* 1: 37-43.

—. 1993. Public uptake of science: a case for institutional reflexivity. *Public Understanding of Science* 2: 321-337.

—. 2003. Seasick on the Third Wave? Subverting the hegemony of propositionalism: response to Collins and Evans. *Social Studies of Science* 33(3): 401-417.

—. 2008. Elephants in the rooms where publics encounter "science"?: A response to Darrin Durant, "Accounting for expertise: Wynne and the autonomy of the lay public". *Public Understanding of Science* 17(1): 21-33.

Ziman, J. 1968. *Public knowledge: an essay concerning the social dimension of science*. Cambridge: Cambridge University Press.

—. 1991. Public understanding of science. *Science, Technology, & Human Values* 16: 99-105.

CHAPTER NINE

THE INCREDIBLE ADVENTURES OF PROFESSOR BRANESTAWM: THE MATURING IMAGE OF SCIENCE IN 20TH CENTURY JUVENILE LITERATURE

ALICE BELL

This paper examines some ways in which images of children or the childlike are used in the construction of the scientist and does so through the study of an interesting series of books from the mid 20th century: Norman Hunter's Professor Branestawm stories (1931-1983). The character of Branestawm, I argue, was clearly constructed as childlike from the start–perhaps as an attempt to appeal to young readers–and yet as the decades move on the character shifts slightly to a representation which, if not overtly "adult" loses the more childish aspects and is juxtaposed in opposition to children. The maturation of Branestawm is not offered to suggest science "came of age" in the latter half of the 20th century, for while this is tempting it is also highly simplistic. Rather, it is an opportunity to explore some of the ways in which images of the childlike can act on cultural constructions of the scientist. I start with a brief discussion concerning images of children and science before introducing the case study. I then outline how I see Branestawm as both a childish and not-childish character before making some further analytical points about how this helps endow the scientist features of the character with some particularly powerful aspects of imagery surrounding the child.

Imagining Children and Scientists

There are several ways in which aspects of the childlike may intersect with the various qualities generally associated with the scientist. It may appear controversial to call a scientist "childlike" as if it was a criticism or form of dismissal. Yet such a link may also endow the image of the scientist with a rather special social status. A childlike science might be seen as trivial, silly or underdeveloped, but it can also be innocent and innately knowledgeable of nature. I have discussed this in more detail elsewhere (Bell 2008a), but it is worth sketching a few points upfront here. We might endow our ideas of children and scientists in ways which make them appear very separate social identities. For example, we might imagine children as stupid, or at least naïve, in comparison to the wisdom of science. Such a figuring may then have an impact on the way we imagine the relationship between the two groups, seeing young people as necessarily recipients of scientific knowledge in need of education to interact with science to become "full" members of society, or for the select few, specific technical training in order to become, after a time, scientists themselves. In contrast, however, we might equally see children as inherently good and connected to nature and juxtapose such imagery to depict science as corrupt and corrupting. This trope is arguably dominant in late 20th century children's science fiction and fantasy, as discussed by Jacqueline Rose (1994), Noga Applebaum (2005, 2006) and Perry Nodleman (1985). There are also ways in which imaginings of the child and scientists might be seen as similar, such as the idea that children are "little scientists" (as expressed, for example, by Lehrer 2011); or conversely–as is the main focus of this essay–that scientists take a positively childish approach to interaction with nature by mixing scientific characteristics of either disinterestedness or curiosity with a sense of childish wonder and naivety.

It is important to stress that at am I not trying to suggest children are one thing or another. Rather, I accept that our ideas of the child are generally multiple and often ambiguously applied–not unlike our ideas of science. In many respects, this essay is less concerned with real children and more interested in the ways our imaginings of children get replicated and reconfigured elsewhere. To assist us in reflecting on the multiple nature of the child in culture, it is worth quoting a rather long passage from sociologist of childhood, Chris Jenks (2005, 2):

> pure, bestial, innocent, corrupt, charged with potential, tabula rasa, or even as we view our adult selves…whether their forms of language, games and conventions are alternative to our own, imitations or crude precursors of our own now outgrown, or simply transitory impenetrable trivia which are amusing to witness and recollect; whether they are constrained and we have achieved freedom, or we have assumed constraint and they are truly free–all these considerations, and more, continue to exercise our theorising about the child in social life.

We can see this in Rousseau and other Romantics, as well as in 19th and 20th century models of developmental psychology, but we should not imagine that such theorising of the child is merely a matter of work in ivory towers. It is important to note that when Jenks refers to "theorising", he not only means academic work, but also the quite prosaic theorising which we all do as part of everyday social life. We all theorise the child as part of social interaction. We draw on others' ideas and/or construct our own based on practical experience and personal moral/political beliefs. Children, unlike other social groups such as those based around professions (for example, the scientist), or issues of race, class, gender and sexuality, have a rather odd status, because we have all been one. The child is then extremely familiar, yet also quite strange as "he or she inhabits our world and yet seems to answer to another; he or she is essentially of ourselves and yet appears to display a systematically different order of being" (Jenks 2005, 2-3; see also Hunt 1994, for an introduction to the politics of child/adult relationships in children's literature).

Similarly, the scientists we are talking about here are very much the "imaginary" ones of fictional depiction. As many sociologists of science have argued, what counts as "science" is a confused, often contradictory notion that is strategically applied depending on context, rather than the consequence of an inherent "demarcation criterion" (Gieryn 1995). In terms of literary representations, we might argue there are right and wrong ways to depict the scientist and complain about incorrect and/or overly negative portrayals (Lambourne 1999). We could, however, acknowledge that there are a range of identities that a scientist in culture might have and a range of reasons why authors would wish to accentuate one or another of them. Taking this latter approach, tracking fictional scientists can become a way of working through the diversity of ways different people at different times have chosen to think about and represent science in society. Rosalind Haynes' (1994) seminal study of the scientist in western literature suggests literary scientists can be sorted into six "stereotypes". Although somewhat reductive, her typology is useful at least as a heuristic and reflects the ways in which scientist characters have in part become cultural tropes, repeated through culture as sketches and allusions. Haynes' types range from alchemists driven to pursue an arcane intellectual goal that carries suggestions of ideological evil to more romantic depictions of the "unfeeling scientist", characters who renege on human relationships in the cause of science. Most important for this study is her sketch of the "stupid virtuoso", a character out of touch with the society and so preoccupied with the trivialities of his private world of science that he ignores social responsibilities. This has similarities to romantic images of the unfeeling scientist, albeit slightly less heroically presented and often comical.

This is not to argue that images of children and scientists remain purely as "make-believe". The fictional, even the fantastical, not only reflect reality (Hume 1984; Hunt and Lenz 2001), but may also have a constructive impact on the doing, making and selling of science. Although scientists may find fiction worrying, seeking to correct its mistakes (Kirby 2003a) or put their own work at a distance (Mellor 2003), they may also find it a useful resource either in the construction of ideas (Kirby 2003b) or for rhetorical positioning (Mellor 2003, 2007). As Megan Prelinger (2010) has shown in terms of discourses of the space race, allusions to science fiction also feature in advertisements for the recruitment of scientists, thereby playing a possible role in the construction at least of some areas of science and engineering. Most interestingly for the purpose of this paper, literary researcher Elizabeth Leane (2007) applies Haynes' typology to those images of scientists constructed autobiographically in non-fiction "popular science" books by scientists, to depict a loose application of Jurdant's (1993) idea of popular science writing as the "autobiography" of science.

My argument, however, will be that all incarnations of Branestawm are at once both childlike and anti-childlike. If we wish to put this in a theoretical framework, I suggest Thomas Gieryn's (1995, 1999) "cartographic metaphor" for the boundaries of science. This considers the cultural entity we call "science" as analogous to a country on a map, its borders defined through processes of boundary disputes and diplomatic treaties. The particular advantage of such "cultural cartography" compared to other frameworks for boundary analysis (for example, Lamont and Virag 2002) is that maps not only show us where two bounded groups are separated, but they also delineate shared space demonstrating simultaneously similarities and distinctions. Gieryn's metaphor also reflects, although in a very different context, Ian Mitroff's (1974) description of scientists' dynamic alternation between "norms" and "counter-norms" of science, as well as what Simon Locke (2005) calls the "ambivalences" of many cultural images of science. As we shall see, Branestawm and, I believe, many "real world" scientists negotiate the child/science boundary to take advantage of both positive qualities of childishness *as well as* the advantages of appearing distinctly "mature".

Introducing Branestawm

The first of Hunter's Branestawm books, *The incredible adventures of Professor Branestawm*, was published in 1933. It comprised of a collection of stories based on those initially read on the BBC Home Service's *Children's hour*. They were written by Norman Hunter, who had previously worked in advertising and as a stage magician as well as serving in World War One in his late teens. He is sometimes known as "Hunter, teller of tales". Indeed, he is referred to as such in bibliographic databases and the British Library copy of the 1933 edition has "teller of tales" sketched in pencil on the frontispiece. The book was illustrated by W. Heath Robinson, who by this time was already famous for his images of fantastic and humorous inventions. *The incredible adventures* was followed in 1937 with *Professor Branestawm's treasure hunt*. After a break of several decades during which the writer lived in South Africa, Hunter returned to writing for children and published fourteen more Branestawm books from 1970 to 1983 (he died in 1995). Although all the Branestawm stories were written by Hunter, a new illustrator was brought in for the 1937 collection and many different artists have worked on the character over the years. Heath Robinson is possibly the most well known, with his images re-used in several subsequent editions (Hunter 1939, 1946, 1993) and so arguably he was a strong influence on the later artists.

Hunter's stories centre on the character of Professor Branestawm: either a scientist or engineer, his specialism ambiguous, who lives in a small English village. Typically the books contain a series of short stories, each of which describes Branestawm inventing something or trying to go about some form of social life, such as attending a party or returning a library book. When conducting his activities, he either gets terribly confused himself or will confuse other people (often both). In most cases, the invention goes wrong and causes some form of havoc or even structural damage. The inventions are in many ways quite magical and often anthropomorphised, and in this respect the books are openly fantastical. It is worth remembering Hunter's work as a conjurer here and that the 1976 spin off activity book, *Professor Branestawm's do-it-yourself handbook*, includes several instructions for "science experiments" that are largely magic tricks (although this is a long-standing trope in such books, including Secord 2002). The experiments might result in blowing up a building or taking over the house. In the 1970 collection, however, *The peculiar triumph of Professor Branestawm*, unusually we find inventions which do work, although not in the way they were supposed to. Interestingly, it is worth noting that the latter stories were "invented" by child readers, who won a competition run by the publisher to design a machine to be featured in one of the stories.

The key feature of the Branestawm character and basis for much of the humour of the books is that he is a social outsider; portrayed as someone the readers and other characters should take some amused bewilderment towards. He is, for many of his fans, the epitome of the absent-minded professor, as he himself reflects at one point: "How strange we professors are, to be sure" (Hunter 1933, 173). Branestawm finds it hard to communicate with people and often has conversations at cross-purposes (Hunter 1933, 130). In one story, he walks into a swimming pool because he is thinking too much about finding a library (Hunter 1933, 88) and he continually forgets where he is. When he discovers the library, he tries to buy a bun thinking it is a bakery (Hunter 1933, 29) and he finds social life difficult, cumbersome and confusing, often feeling out of place and lost. The playfulness of this stereotype is very explicit in one story whereby he ends up inadvertently doing an impersonation of himself at the seaside, which results in people thinking he is a comic pretending to be an absent-minded professor (Hunter 1933, 203). Branestawm has a housekeeper, Mrs Flittersnoop and a best friend, Colonel Deadshott, but he generally chooses to be left alone to work with his "inventions".

The books have been translated into several languages, including Polish, Swedish, Italian, German and Thai.[1] There was a television series in the late 1960s and later a BBC radio adaptation in 2001. One study of books used in British schools in the 1970s stated they were amongst the top 20 books read aloud in primary schools, alongside the *Chronicles of Narnia, The Borrowers* and various works by Dahl, even beating the *Just William* stories (Poole 1986). They are still in print in the UK, although this is arguably largely due to the nostalgic tendencies of children's publishing as the books are less common today. Arguably, this is due to changes in social attitudes towards race, gender and class in the 1980s, as the way they are represented in the books became disagreeable, which would have kept them off school bookshelves from then onwards. In places, however, Branestawm appears to continue to resonate in British cultural imagery of the scientist and people might refer to academics as having "Professor Branestawm hair" as if it is an example everyone will know. Christopher Frayling (2005, 221) suggests that the shorthand sketches children compose when asked to draw a picture of a scientist may come from computers, comics, cartoons, films or the cover of Professor Branestawm books. Notably, Frayling refers broadly to each medium except for books where a very specific, if rather out-dated example is given. However, this only shows the power and apparent universality of the Branestawm image in particular cultures to which Frayling appears to belong.[2]

The Variable Childishness of Professor Branstawm

I believe the original 1933 Branestawm to be quite overtly constructed as childlike, both in Hunter's text and Robinson's illustrations. Yet, Branestawm's childishness is variable across the 20th century. As with any long-standing character, such as Dr Who, Superman, or Tom Swift, the cultural referents used to construct the character and stories change over time depending on the context. However, Branstawm retained the same author, albeit with changes in illustrators and editors. One of the more overt examples of how shifts in the public views of science had an impact on changes in the book are James Adams' illustrations in the 1937 edition, as these are much more heavily influenced by images of Freud and depict a hairier, bearded Branestawm in a neatly drawn pinstriped suit. As I will argue, the character–more in text than illustration–seems to "mature" somewhat when Hunter returns to writing about him in 1970, becoming less and less childlike throughout the decade, before reaching a sort of "second childhood" in the 1980s incarnations. This section will therefore provide a sketch of some of the more childlike aspects of (the generally early) Branestawm, sometimes in comparison with (later) more apparently "mature" forms of the character.

To start with Branestawm's childishness, for all his occasional seriousness in the face of social events such as parties, he clearly has a sense of fun and when the fair visits town for example, he is desperate to go (Hunter 1933, 69ff.). There are also childlike references to eating too much trifle and drinking too much ginger beer (Hunter 1933, 86), while at bedtime he drinks cocoa and falls out of bed (Hunter 1933, 169). His lack of social understanding also leads to quite childlike concerns and he takes things to the extreme. For example, at the end of a fancy-dress party, where the Professor and Colonel have gone as each other, people do not believe they had changed back from fancy-dress: "So they very nearly had to live in each others' houses for the rest of their lives" (Hunter 1933, 154). The early illustrations show both Branestawm and the Colonel as reasonably short compared to other characters and they often crouch over slightly, without the more confident posture of some adults. In one illustration Branestawm is shown wearing a "trouser elevator contraption" (Hunter 1933, 3), which is a set of braces attached to a fishing rod mechanism and stops baggy clothes falling down. This also has the effect of bunching the trousers around his bottom making him look as if he is wearing a nappy. In the Heath Robinson images of Branestawm, he is often shown wearing clothes that are too big for him, which is indicative that he was playing at fancy dress and not quite performing an adult role. The picture in the 1933 frontispiece shows Branestawm's tie worn back to front, which is there as a symbol of the lack of care he gives to his clothes and a function of his absent-mindedness, but is also reminiscent of an untidy schoolboy.

On some occasions the childlike is, perhaps oddly, constructed through references and symbols of old age. For example, he is depicted as balding. A baldhead in many ways symbolises old age, however it also has the effect of making the character look as if he had not grown any hair yet. We are also told he has a large head because he is so brainy (Hunter 1933, 7), but again a large forehead is a rather baby-like characteristic. Interestingly, this is perhaps most overtly underlined when we see an image of Branstawm compared to himself as a baby. The Professor invents a potion that brings images to life but his housekeeper accidentally pours it over a photo album and facsimiles of Branstawm through the ages run amok throughout the house. One image shows Branestawm taken aback by himself as a baby (Hunter 1933, 121). The baby sits serenely on the floor, on the left hand side of the page, grabbing his feet. The Professor attempts to mirror this posture and falls back in alarm whilst bending his body in the middle to reflect the baby's movement out to touch his toes. Although the Professor is much taller than the baby and clothed, the differences largely end there. They are both generally hairless. Significantly, the baby's hair is the coarse and frizzy type which is more characteristic of an ageing person than the light and soft curls of a baby. Further, the baby wears glasses despite being clearly too young to wear them. To me this seems like a clear example of "ambivalence" around child/adult boundaries with respect to a scientist character, as it draws attention to similarity and difference at once. I also think it shows that Heath Robinson picked up on the same cues in the text as I spotted.

The 1930s Branestawm is seemingly terrified of people. Characters from the rest of the village, when we see them, are shown to like him, but generally he hides from social life with the exception of one friend, Colonel Deadshott. Such anti-social behaviour is also in some ways quite childlike, reflecting a child's lack of socialisation or shyness, the distinction being that an absent-minded professor is innocent of society by choice or inclination; for a child it may be more temporal, but the effects are similar. Branestawm's lack of socialisation is also symbolised through references to his odd approach to clothes, language and food. Again, these could be seen as childlike, having not been learnt. These traits are humorous because they are incongruous to what might be expected as normal, but there are also ways in which they figure the character as cute, almost helpless in a childlike way. This characterisation is underlined further by the housekeeper who is there to take care of him. However, they also make him slightly inhuman, almost alien. With respect to food, he would have breakfast overnight so he could start on an important job earlier the next morning (Hunter 1933, 15). With clothes, as well as the fancy-dress and trouser contraption incidents mentioned earlier, he is referred to as making notes on his shirt cuffs (Hunter 1933, 16) and at one point inadvertently puts on a lady's bonnet (Hunter 1933, 48, illustration). Hunter obviously enjoyed linguistic play and although notes such as, "He had very few friends because people found it so very difficult to talk to him" (Hunter 1933, 1) are perhaps a nod to scientific jargon, the language used by the character is more a matter of being vague and imprecise, using phrases such as "thingummy" and "whateveritis" (Hunter 1933, 3) that suggest a lack of familiarity with language rather than having developed his own advanced form of it. Perhaps most significantly, people are indulgent or at least tolerant of his odd behaviour and Deadshott pretends to understand him (Hunter 1933, 4). In one incident, he blows a man's house up so the Professor offers to build him a new one, "But the clock-man said he'd rather invent his own house, thank you all the same" (Hunter 1933, 66).

In the 1930s, Branestawm hid away in his own private space of the home, but by the 1970s we see him out and about interacting in public spaces such as libraries, schools and parks. In the 1980s, we see a sort of synthesis of the two locations as the public world is brought into the home. It is thus a very domestic setting, but one at home and familiar with the outside world. In the 1980s, he becomes much more explicitly affable and loveably foolish again: he has parties with jelly and ice-cream (Hunter 1983b); and writes letters (Hunter 1981b). However, although he interacts with the world in a familiar way and is not as seemingly terrified as he was in the early books, he generally stays at home and is pleased when guests go home, producing a machine because he does not like writing thank yous. He does not want people around, although now it does not seem as if it is because he is scared of others, more that he has learned they are annoying.

In the 1970 book (Hunter 1970), there are also several encounters with children emphasising the distinction between the two social roles, but with less of the juxtaposition to show similarity seen in the 1933 Heath Robinson illustrations. As Branestawm moves into the 1970s, the central characters of these books are now intended for the children to laugh at and look at from afar, rather than play at being. They are funny because they are odd *to* the children, not odd *like* the children. Whereas the 1930s inventions tend to be based in Branestawm's home, designed to solve problems about running the house, in the 1970s inventions are on a larger scale and located outdoors in the centre of social life in parks, shopping streets and schools, working for the public good of the whole village not just himself. In the 1974 book, *Professor Branestawm's great revolution*, the last story reveals what the revolution in the title refers to when the Professor is asked to design a revolving restaurant. He also lectures at an institute for science and other scientist characters are brought in. There are references to the links between science and the military, as well as commercial culture. There is also a greater sense of specialisation in contrast to the first book, where it is not clear what he was a professor of as he invented machines suggesting he was an engineer, but also wrote books on languages. Post-war, he is more explicitly fashioned as a scientist/technologist, as when he goes abroad, for example, not to study people anthropologically but to learn more of their inventions (Hunter 1970, 37). Looking at the language, the Professor also discusses the science in slightly more detail, suggesting a rather more "adult" vocabulary compared with the "thingummy" and "whateveritis" of the first book. There is even a rival Professor, Gasket Basket, who he gets into an inventing competition with, but who uses underhand methods such as employing "rascals" to steal Branestawm's invention. Branestawm's inventions in the 1980s books, however, return to rather more domestic settings (Hunter 1981a, 1981b, 1982a, 1982b, 1983a, 1983b).

The Rhetoric of the Childish Scientist

There is a sense that the inability to interact with society is a necessary consequence for scientific achievement and, by that vein, also socially acceptable:

> For although the Professor was so clever, or perhaps *because* he was so clever, he was very absent-minded. He was so busy thinking of wonderful things like new diseases or new moons that he simply hadn't the time to think of ordinary things like old spectacles. (Hunter 1933, 1)

This characterisation is very similar to what Rosalind Haynes (1994) calls the "stupid virtuoso" and arguably pre-dates the more Victorian image of childhood represented in the book. Certainly, an absent-minded professor is not unique to these books, however there are ways in which stupid virtuoso characters can be endowed with aspects of childishness on which it is worth reflecting, to deconstruct them and consider the ways they may influence our ideas about children's relationships with science (and vice versa). From a study of Richard Feynman's popular writing, Leane (2007, 151) argues that behind anecdotes of asking for lemon *and* cream in your tea, or asking a servant to dance, lies a notion of "the wise fool", or as she puts it, "the boy who saw the emperor's nakedness". These stories are applied to show how the scientist can debunk unnecessary or cruel social practices and find the efficient, true and (morally) right way ahead. Leane does not reflect on it being a boy rather than a person of any age who manages to expose the "nakedness", but this unconscious reference to youth is significant as it concurs with the idea of the innocent child intersecting with notions of the scientist as either "disinterested" (Merton 1973), or a "modest witness" of nature (Shapin and Schaffer 1985).

Linked to this idea of innocence is a strong sense of simpleness in the image of Branestawm, which sits within some of the figurings of him as having a childish character whilst not being dependent on it. For example, again from the start of the first book (Hunter 1933, 1):

> Professor Branestawm, like all great men, had simple tastes. He wore simple trousers with two simple legs. His coat was simply fastened with safety pins because the buttons had simply fallen off. His head was simply bald and it simply shone like anything whenever the light caught it.

This equating of an idea of simplicity to both science and greatness reflects cultural ideas of the scientist as humble as well as more philosophical views, such as Occam's razor. It also implies we need not question him: he just is.

Yet, interestingly, behind the appearance of simplicity lies a more complex sense of magic as the characterisation also plays up to the idea of the scientist having an intuitive or hidden ability to work nature, for example in the description of one of Branestawm's inventions, which is pretty standard for many throughout the whole series until the 1980s:

> Blue and yellow smoke shot out from every part of the machine. Wheels whizzed. Levers clicked. Little bits of stuff went buzzing up and down and round and round. And far beneath them the landscape rushed by quicker and quicker until at last they could see nothing but a grey haze all around them. (Hunter 1933, 4-5)

Similarly, from 1981 (Hunter 1981b, 5): "Suddenly he had an idea. Right out of nowhere it came. Like catching measles, only nicer." It is maybe worth remembering Hunter's time as a stage conjurer who fancied himself in the fancy-dress of the mad scientist, an odd reflection of J.B.S. Haldane (1937) perhaps, who in his Mr Leakey stories adopts the costume of a magic man, but equally uses a children's book to play with ideas of the scientist as magician. As Haynes' history of fictional scientists shows, this idea of scientist as magic man has a long history and if anything it is an especially old form of the character. However, it is also worth noting that this almost magical connection between the child and nature is a very common theme in children's literature–as Rose (1994, 43) puts it, the child is "constantly set up as the site of a lost truth". Applebaum (2006) has argued that such nature-child connections are used to distance children from science and critique the latter in many science fiction books written for young people. However, I think Branestawm is an example of the ways in which the two can be combined, endowing the scientist character with the various positive characteristics of the child mysteriously connected to nature (for other examples of this argument, see Bell 2008a, 2008b).

Concluding Points

This essay has considered some of the ways in which an image of the child might be woven into that of a scientist. I hope this helps us consider some of the detail of our cultural construction of scientists, in particular their rhetorical appeal. Oddly perhaps, the childlike can endow the image of the scientist with a rather unthreatening and intuitive association with natural knowledge, even if childlike scientists will also have to work to distance themselves from other images of youth such as the more pathological notions of ignorance, triviality or thoughtlessness. As Elizabeth Leane's (2007) discussion of the constructed public face of Richard Feynmen suggests, there is a rhetorical appeal to such constructions as childlike imagery may provide powerful symbols of an apparent link to truth and/or purity of purpose. Leane's study also shows how these fictional representations spill over into non-fictional ones. Branestawm is not part of a concerted public relations campaign for science, but such images can be extended elsewhere to become so.

It would be a neat piece of analysis to argue that the differences in age-based allusions of Branestawm are reflective of the emergence of "Big Science" as a form of coming of age, but that would be too simple. To some extent this analysis is ahistorical anyway, or rather, cross-historical, not because what Branestawm depicts is in any way universal, but because children's literature is so often carried across time. Due to the nostalgic tendencies of children's book production and dissemination (i.e., that it is made by the generation before its audience and parents often give books from their youth to their children), what Branestawm said about science then is largely what it says now. Branestawm is thus interesting today, not only as an expression of science in the mid 20th century, but one that both looked back nostalgically at the time and remains resonant in its nostalgia today, either through various recyclings of postmodern culture or more directly as readers of the 1970s give it to their children, who in turn give it to theirs. Instead, the variability and contingency of Branestawm's symbolic childishness perhaps provides an extra analytical point. In the latter half of the 20th century, scientist characters became more central to public life, but at the same time would emphasise their special qualities to appear distinct. If I was to hazard any broad statement about the sweep of time the books reflect, it is that images of science became more ambivalent, drawing on the childlike images of the past while adding in new features. And Branestawm may yet return, perhaps further re-fashioned by a new author, illustrator or production team. Although the most recent edition of the 1933 book (which is the most long-lived) I could find was published a decade ago, there are e-book versions, while the BBC last re-broadcast a reading of the stories in March 2011 and an online bookshop listing suggests a "Vintage" edition will be published in 2013.

Notes

1. My thanks to the "academic discussion of children's literature–uk" listserv for assistance in tracking down Branestawm in translation.
2. It should be noted that Frayling is discussing a set of drawings produced by 7-11 year olds in the South West of England in summer 2003. An argument for the resonance of Branestawm in the subjects of the 1983 UK study he is replicating might be more convincing.

Acknowledgements

This work was carried out as part of doctorial research (AHRC grant no: 05/116470) at the Science Communication Group, Imperial College. Drafts of this paper were presented at several conferences. My thanks to the organisers of these events, as well as all attendees for their attention, suggestions and questions.

References

Applebaum, Noga. 2005. Electronic texts and adolescent agency: computers and the internet in contemporary children's fiction. In *Modern children's literature: an introduction*, ed. Kimberley Reynolds, 250-263. Basingstoke: Palgrave Macmillan.

—. 2006. The myth of the innocent child: the interplay between nature, humanity and technology in contemporary children's science fiction. *The Journal of Children's Literature Studies* 3(2): 1-17.

Bell, Alice. 2008a. The childish nature of science: exploring the child/science relationship in popular non-fiction. In *Science and its publics: following scientists into popular culture*, eds. Alice Bell, Sarah Davies and Felicity Mellor, 79-98. Newcastle: Cambridge Scholars Publishing.

—. 2008b. Science as pantomime: explorations in contemporary children's non-fiction books. PhD diss., Imperial College London.

Frayling, Christopher. 2005. *Mad, bad and dangerous?: The scientist and the cinema*. London: Reaktion Books.

Gieryn, Thomas F. 1995. Boundaries of science. In *Handbook of science and technology studies*, eds. Sheila Jasanoff, Gerald Markle, James C. Petersen and Trevor Pinch, 393-443. London: Sage.

—. 1999. *Cultural boundaries of science: credibility on the line*. Chicago: University of Chicago Press.

Haldane, J.B.S. 1937. *My friend Mr Leakey*. Harmondsworth: Penguin.

Haynes, Rosalind. 1994. *From Faust to Strangelove: representations of the scientist in western literature*. Baltimore: Johns Hopkins University Press.

Hume, Kathryn. 1984. *Fantasy and mimesis: responses to reality in western literature*. London: Methuen.

Hunt, Peter. 1994. *An introduction to children's literature*. Oxford: Opus.

Hunt, Peter, and Millicent Lenz. 2001. *Alternative worlds in fantasy fiction*. London: Continuum.

Hunter, Norman. 1933. *The incredible adventures of Professor Branestawm*. Illustrated by W. Heath Robinson. London: John Lane.

—. 1937. *Professor Branestawm's treasure hunt*. Illustrated by James Arnold. London: John Lane.

—. 1939. *Stories of Professor Branestawm*. Illustrated by W. Heath Robinson. Leeds: E.J. Arnold & Son.

—. 1946. *The incredible adventures of Professor Branestawm*. Illustrated by W. Heath Robinson. Harmondsworth: Penguin.

—. 1970. *The peculiar triumph of Professor Branestawm*. Illustrated by George Adamson. London: Bodley Head.
—. 1974. *Professor Branestawm's great revolution and other incredible adventures*. Illustrated by Derek Hughes. London: Bodley Head.
—. 1976. *Professor Branestawm's do-it-yourself handbook*. London: Bodley Head.
—. 1981a. *Professor Branestawm's pocket motor car*. Illustrated by Gerald Rose. Harmondsworth: Puffin.
—. 1981b. *Professor Branestawm and the wild letters*. Illustrated by Gerald Rose. London: Bodley Head.
—. 1982a. *Professor Branestawm's mouse war*. Illustrated by Gerald Rose. London: Bodley Head.
—. 1982b. *Professor Branestawm's building bust-up*. Illustrated by Gerald Rose. London: Bodley Head.
—. 1983a. *Professor Branestawm's hair-raising idea*. Illustrated by Gerald Rose. London: Bodley Head.
—. 1983b. *Professor Branestawm's crunchy crockery*. Illustrated by Gerald Rose. London: Bodley Head.
—. 1993. *The incredible adventures of Professor Branestawm*. Illustrated by W. Heath Robinson. London: Puffin.
Jenks, Chris. 2005. *Childhood*. Second edition. Oxford: Routledge.
Jurdant, Baudouin. 1993. The popularization of science as the autobiography of science. *Public Understanding of Science* 2: 365-373.
Kirby, David. 2003a. Science consultants, fictional films, and scientific practice. *Social Studies of Science* 33(2): 231-268.
—. 2003b. Scientists on the set: science consultants and the communication of science in visual fiction. *Public Understanding of Science* 12(3): 261-278.
Lambourne, Robert. 1999. Science fiction and the communication of science. In *Communicating science: contexts and channels*, eds. Eileen Scanlon, Elizabeth Whitelegg and Simeon J. Yates, 146-157. London: Routledge.
Lamont, Michele, and Molnar Virag. 2002. The study of boundaries in the social sciences. *Annual Review of Sociology* 28: 167-195.
Leane, Elizabeth. 2007. *Reading popular physics: disciplinary skirmishes and textual strategies*. Aldershot: Ashgate.
Lehrer, Jonah. 2011. Every child is a scientist. *Frontal Cortex: Wired Science Blogs*, September 28. http://www.wired.com/wiredscience/2011/09/little-kids-are-natural-scientists/

Locke, Simon. 2005. Fantastically reasonable: ambivalence in the representation of science and technology in super-hero comics. *Public Understanding of Science* 14(1): 25-46.

Mellor, Felicity. 2003. Between fact and fiction: demarcating science from non-science in popular physics books. *Social Studies of Science* 33(4): 509-538.

—. 2007. Colliding worlds: asteroid research and the legitimatization of war in space. *Social Studies of Science* 37(4): 499-531.

Merton, Robert K. 1973. *The sociology of science: theoretical and empirical investigations,* ed. Norman W. Storer. Chicago: University of Chicago Press.

Mitroff, Ian I. 1974. Norms and counter-norms in a select group of Apollo moon scientists: a case study in the ambivalence of scientists. *American Sociological Review* 39: 579-595.

Nodleman, Perry. 1985. Out there in children's science fiction: forward into the past. *Science Fiction Studies* 12: 285-295.

Poole, Roger. 1986. The books teachers use. *Children's Literature in Education* 17(3): 159-180.

Prelinger, Megan. 2010. *Another science fiction: advertising the space race 1957-1962*. New York: Blast Books.

Rose, Jacqueline. 1994. *The case of Peter Pan*. Second edition. Basingstoke: Macmillan.

Secord, James A. 2002. Portraits of science: quick and magical shaper of science. *Science,* September 6: 1648-1649.

Shapin, Steven, and Simon Schaffer. 1985. *Leviathan and the air-pump: Hobbes, Boyle, and the experimental life*. Princeton: Princeton University Press.

CHAPTER TEN

HOW POPULAR CULTURE ENGAGES AND DEBATES SCIENTIFIC THOUGHT: SCIENTIFIC AND SUPERNATURAL NARRATIVES IN LATE VICTORIAN GOTHIC

KATHERINE ROACH

Fictional representations of science and scientists form a substantial proportion of the contact that non-scientists have with science and its practitioners. Yet with notable exceptions, the body of work on science in fiction tends toward too simple an account of the way in which readers interact with fiction. Much as Adorno and Horkheimer (1997 [1944]) argued in 1944 of a mass culture absorbed wholesale by the populace turning their fighting spirit to docile acquiescence, so studies of science in fiction have had a tendency to assume a passive audience that soaks up meaning from texts in a direct and naive way. There has also been an emphasis on a single, frightening image of science, often referred to as the "mad scientist", which is too readily taken as a straightforward expression of negativity and fear of science indiscriminately mopped up by the masses.

Here, I will begin with the observation that in the early 20th century, *the* towering figure of mad science, Mary Shelley's Frankenstein, was continually paired with another scientist of equal and opposite status–the occult scientist, vampire hunter and Dracula's destroyer, Professor Van Helsing of Bram Stoker's *Dracula* (1997 [1897]). Stoker's scientist gets short shrift in studies of science in fiction probably because supernatural investigation becomes science's "other" in 20th and 21st century terms.

Indeed, by placing the mad scientist as a (rightful) example of Alfred Schutz's (1967) typification, then it becomes clear that the lack of attention paid to occult science arises because the occult is no longer a typical object of science as it was in the 19th century. Further, Schutz's phenomenological turn places fictional images as part of the warp and weft of social knowledge, which we use to create meaning and to make sense of the world. So the question shifts away from the naïve, "what does fiction do *to* people" to, "what does fiction do *for* people".

Having established that these stock characters bubble up from social knowledge, it is clear that they are not simply to be enjoyed. They serve a purpose as cultural tools with which we can debate and think through the meanings of science in society (an issue not reserved for academics). The continual pairing of *Dracula* with *Frankenstein* engages a compelling discourse about the disenchanting (or not) effects of science and rationalism on the social world. When paired, these two narratives encompass the enchanted and the disenchanted and as such have come to mean more together than they do alone.

Moving on, I explore the vampire hunter in more depth and show that Professor Van Helsing is *the* abiding example of an "occult detective", the one remaining model from a number of examples of similar characters who tested the boundaries of the natural and supernatural in the late 19th century. Where did enchantment end and disenchantment begin? The question was a focus of frenzied activity across several domains of fin de siècle culture and, as part and parcel of the social world, fiction offered its own brand of tools for thinking it through, in ways that I go on to explore.

A notable fictional forerunner of the occult detectives was Dr Hesselius from Sheridan Le Fanu's (1993 [1872]) short story collection *In a glass darkly*. Despite being an inspiration for Stoker's *Dracula*, Le Fanu's tales are sophisticated reflections on the authority of rationalism, which is never given complete ascendancy. The line between enchantment and disenchantment is left undrawn and the question unresolved. In comparison, Stoker's *Dracula* takes on the narrative arc of a detective story, which resolves in the detection and destruction of the vampire.

There are synergies between the narrative structures of science itself and detective fiction, and between detective fiction, science and the way in which readers attend to the chronology of fictional narratives more generally. Science or detection tells a good story because it reflects patterns of narrative processing that non-scientific readers employ to understand stories. All are reliant upon the dependable chronological passage of time. Causes are followed by effects and not vice versa.

Yet the supernatural disrupts the natural passage of time. Ghosts, ghouls and vampires constitute an invasion of the past into the present. The rational investigation of supernatural causes destabilises the scientific turn of mind. When time is disordered, cause cannot be relied upon to precede effect–with one plausible exception that is: the vampire has a material body and materiality allows for a seeming, if superficial, process of rational cause and effect. Although ancient, capable of shape-shifting and all manner of magical doings, the vampire's body lends a veneer of plausibility to empirical detection. The body can be hunted down whether it is enchanted or not. Bram Stoker's story manages the impossible in combining the cogency of cause and effect with a dash of belief in mercurial magic. The mix allows Professor Van Helsing to succeed in vanquishing evil where Victor Frankenstein disregards the supernatural and fails. The trouble with evil is that it always returns, so what is needed next time? Will it be science or will it be magic? Quite a hefty debate for such low-brow fictions.

Frankenstein's Foil

The mad scientist is an icon of popular culture as familiar as a cup of tea or a can of soda, common enough in fact to have an entry in the *Oxford English dictionary*.[1] He, for male he usually is, shares a narrative with Mary Shelley's (1994 [1818]) character, Victor Frankenstein. An ambitious scientist turned misanthrope manufactures a man, bomb or other being which disastrously runs out of control. The ubiquity of the image is a likely reason for the strong emphasis that studies of science in fiction have placed upon mad, bad and dangerous types. The import of the icon is further underlined by a slow but steady stream of writing that flows from the scientific community bemoaning a perceived dearth of decent scientists in popular culture (Baldick 1987; Basalla 1976; Cohen 1981; Frayling 2005, 2006; Haste 1997; Haynes 1994, 2003, 2006; Jackson 2008; Jones 2001; Jörg 2003; Millhauser 1973; Rose 2003; Schummer 2006; Skal 1998; Toumey 1992; Tudor 1989; Turney 1998; Weart 1988; Weingart 2006; Weingart *et al.* 2003; Weingart and Pansegrau 2003; and for scientists on fiction, see Emsley 2001; Sir David King in Frayling 2006; Gilbey 2008; McDonald 1989; Pollack 1998; Rohn 2006).

Each seam of writing supports the other in a near exclusive focus on an unruly, unkempt and unsafe scientist. A narrow vision indeed, given that almost no part of our lives, in the west at least is left untouched by science and technology, from food and shelter to health and communication. It would then be extraordinary if the sum total of our imaginative response to science were concentrated in such a fearfully negative image.

Indeed, this is not the case. There are other scientists who claim equal status in the ancestry of fictional science. In the 20th century, *Frankenstein* gained a partner in *Dracula*. Although *Dracula* is not reckoned to be a story about science in an obvious way as *Frankenstein* perhaps is, the novel has a powerful scientific protagonist. Where Victor Frankenstein creates a monster, Stoker's Professor Van Helsing destroys one. Bram Stoker's (1997 [1897], 147) story tells of a vampire hunt in which Van Helsing, "a philosopher, a metaphysician, and one of the most advanced scientists of his day" saves humanity from the curse of the "undead" by injecting a shot of calm empiricism into the mission to lay such dangerously disturbed souls to rest.

A now time-honoured affiliation between the two 19th century horror stories had humble beginnings in December 1927, in twin stage adaptations of Hamilton Deane's repertory company, which toured the UK provinces bringing double demonology to a West End debut in 1930. Soon afterwards, Universal Studios famously took an interest and the two great horror icons drifted into popular culture as a matching set in horror double bills, compendium story books, comics, television shows, games and all manner of cultural products (Skal 1993). Few on-screen scientists can have rivaled Victor Frankenstein for air-time more than Van Helsing has in the 20th century. Jon Turney (1998) finds over 40 film adaptations of Frankenstein up to 1982 and in Stephen Jones' (1993) movie guide compiled a decade later, he counts 400 films loosely based on the same story. A quick count of movies listed by the same author reveals that there are 664 vampire films made before 1990 that include a character based on Van Helsing.

Such a profusion of products arising from the demonic duo exemplifies the kind of economically driven cultural homogenisation that so appalled Theodor Adorno and Max Horkheimer when they wrote about the culture industry. In the simplest of terms, their argument, the bones of which Adorno (1975) reiterates, hinged on the fact that "art" whether literary, musical or visual had become a saleable commodity and as such was no longer created in the purpose of stirring the conscience, but simply as a product to be "enjoyed". Instead of galvanising the dissatisfied to fight for their cause or at least to reflect upon it as art could, the culture industry delivers an escapist activity that supports rather than questions the status quo and replaces "consciousness" with "conformity" among the masses (Adorno 1975, 17).

This element of the concept of a culture industry has echoes in discourses past. Chartists, radical writers and pamphleteers from the 1840's onwards were generally suspicious of burgeoning weekly "penny fictions", because they believed that they offered up imaginary answers for genuine social problems and were a distraction from the real work of social change. Morag Shiach (1989) has described two poles of response to "popular" culture in this period: at one end she finds this radical view;

at the other end were educators, reformers and religious campaigners who argued that cheap fictions were a bad influence on the morals of the masses. Hepworth Dixon, writing in the *Daily News* in 1847, felt that they could

> poison the very fountains of human life…by corrupting…the basis of morals…[that] offer the only guarantee for the peace and well-being of the social body (cited in Dalziel 1957, 49).

By these accounts, then, popular fiction is rabble rouser or soporific dupe. Shiach's (1989) study suggests that debates about the "popular" in culture have long been defined by these two poles. A sophisticated tradition of theorising the popular extending from the 19th century to today spans both extremes of opinion, but is generally unified in describing the popular variously as de-based, dumbed down, formulaic, low-brow, vulgar and unvarying. Adorno is as damning as they come, yet he, unlike many of the 19th century writers on the topic, did engage with the material of popular culture, for he wrote extensively about "popular music". He replied to the reprimand, "Whoever ignores [the culture industry's] …influence out of skepticism for what it stuffs into people, would be naïve" (Adorno 1975, 15). He sweeps away any call to take it seriously as a distraction from the true critical focus on quality and truth. Whatever the value of his position, it constructs the cultural consumer as entirely passive, unresponsive and waiting to be "stuffed" with undemanding material.

A similar trend runs tacitly through much writing about science in fiction, which emphasises the mad scientist as *the* standard image and then bemoans the fact that it dulls lay understanding of science and stops genuine debate or learning dead in the water (Frayling 2005, 2006; Haynes 1994, 2003, 2006; Kirby 2003; Millhauser 1973; Schummer 2006; Toumey 1992; Turney 1998; Weart 1988; Weingart 2006; Weingart *et al.* 2003; Weingart and Pansegrau 2003; for scientists making the same claim see Emsley 2001; Gilbey 2008). True, mad scientists are unreal, but they are designed that way. They are the stock characters of fiction and as such they have little psychology, little history and little personalising detail at all. In this they are much like the characters of a fairy tale.[2]

Some scholars have argued that the characters themselves constitute an unbending critique of science. Most mad scientists have strange equipment, sometimes galvanic, often bubbling vials or other arcane instruments and they are commonly unkempt, male and wear white coats.[3] Christopher P. Toumey (1992) argues that the paraphernalia of mad science consists of an illogical bundle of magical iconography that amounts to an antirational critique. Other scholars focusing on the disaster narrative itself are inclined to take that as a straightforward reflection of fear or negativity on the part of public(s) (Basalla 1976; Cohen 1981; Frayling 2005, 2006; Gilbey 2008; Haste 1997; Haynes 1994, 2003, 2006; Kramer 2003; Millhauser 1973; Pollack 1998; Rose 2003; Weart 1988; Weingart 2006; Weingart *et al.* 2003; Weingart and Pansegrau 2003). Yet mad science does not necessarily amount to "bad" science. Stories of mad science have been documented as rhetorical devices used by pro-embryo research groups to suggest that opponents to embryo research were confusing fact and fantasy (Mulkay 1995, 1996). Lindy Orthia (2010) has also described how the popular television series *Dr Who* turns classically drawn scientific villains into positively pro-science messages, when the Doctor's science outshines the villain's.

Perhaps what is called for is a shift in the framing of debate away from "what does popular fiction (mad science stories) do *to* the people?" Is the story of a crackpot man and his misanthropic unfettered ambition really taken as read by a populace who cannot tell the difference between a stock character of fiction and the workings of real-world science? What happens if we ask instead, "what does popular fiction do *for* the people?" Why are the twin demons of Frankenstein and Dracula so commonly paired? What do these two figures do *for* people? Do they really dupe us into delightful dreams so that we are distracted from real-world problems that need our attention? To some extent of course they may well do. It is hard not to have some sympathy with the concept of the culture industry particularly in the current climate of consumerism that thrives alongside global warming and recession. It would be going too far to suggest that the culture industry has no relevance, but it seems unlikely that we replicate these figures so avidly *only* for their anesthetic value.

Recent work does suggest that such fictional icons work for people in different ways. Focus group participants used them as framing devices for their discussion about the impacts of science on society (Bates 2005; Michael and Carter 2001). When asked about the representations themselves, participants referred to them creatively and intelligently, and easily assimilated different kinds of knowledge from different contexts that spanned didactic, entertaining, emotional, ethical and rhetorical purposes. In other words, they treat fiction as symbolic and not as "real".

Images from fiction, stock characters or otherwise, can be utilised in extraordinarily complex, multilayered ways, which suggests that though they might provide an entertaining distraction they also contribute to a pool of cultural resources for thinking and debating about society and science (Locke 2005). In this view, Frankenstein, Dracula, the mad scientist and the vampire hunter are iconic cultural reference points drawn into the everyday social world and made to do work for us as framing devices, as rhetorical devices, as material products of a capitalist system or as simple entertainment.

Beyond Typical Science

Bram Stoker's Professor Van Helsing is one of a number of scientific "occult detectives" that appeared in the latter third of the 19th century (e.g., Le Fanu 1993 [1872]; Blackwood 2002 [1908]; Hodgson 2006 [1913]; Heron and Heron 2006 [1899]). They have been little featured in writing on science in fiction and yet they are scientists through and through–symbols perhaps of hidden or arcane knowledge (including science) and the very epitome of rationalism, with the rare ability to keep a "straight head" in the face of chaos and annihilation. The reverse, indeed, of the mad scientist who loses his head to his ambition. The occult detectives, of which Frankenstein's Hollywood partner Van Helsing is probably the most famous, have a strong presence today and still they are practically ignored by scholars of science in fiction. One reason for the seeming lack of interest might be a tautological knot of sorts. Though there is a call for decent scientists in fiction, those who are decent, the occult scientists, may not be recognised as scientists because they dabble in the supernatural, do not spend their days labouring in a laboratory and lack other typified emblems of science.

"Typified emblems" or the typification to which I refer is a concept that has phenomenological underpinnings and was initially theorised by Alfred Schutz (1967). For him, individuals understand the world by interpreting it in terms of a stock of previous experience, which encompasses their own experiences combined with that communicated by parents, relatives, friends and cohorts. This stock knowledge forms a schema through which all new experience is measured, so that each new object and event is perceived at the outset in a typical character–a typification. In this way, the world can take on meaning without the necessity to process each new experience from afresh. Typifications are essentially learned common sense constructs applied to new sensory experiences of places, people, events and things. Schutz argues that it is through this process of typification that individuals are able to take meaning from situations of which they have no personal experience. He (1973, 17) explains

> [p]utting a letter in the mailbox, I expect that unknown people, called postmen, will act in a typical way, not quite intelligible to me, with the result that my letter will reach the addressee within typically reasonable time.

Schutz (1973, 12) proposes a "general thesis of reciprocal perspectives" to explain how people are able to bridge the social-subjective divide. It goes something like this: for me to be able to assume that the world that I know as real is the same as the world that you take for granted is real, then we must share the same taken-for-granted thinking structures about the world. In fact, everyone within our social group must have access to the same stock of knowledge with which we can move from our own subjectively constructed worlds to our socially constructed world. This must be socially agreed. If this does not happen then we are thrown into our individual solipsistic worlds.

In order to engage in the same reciprocal arrangement with regard to a single entity that we wish to refer to as "science", then we must assume common sense constructs that are able to encompass an enormously diverse group of activities and people from zoologists to astrophysicists. We must develop a stock of thought constructs that allow us to refer to the essential characteristics of what a "scientist" does. Such constructs will encompass the "sedimented" products of forms past, which have been stretched, added to and lost details over time. This is the process of typification. It reflects the way that 20th century insignias of Einstein's hair and a white coat have been added to 19th century detail of bubbling vials and galvanic apparatus as typified emblems of today's mad science.

Like shifting sands, typifications of the 20th and 21st centuries do not quite match those of the 19th century such that today's scholars do not share the socially sanctioned knowledge of 19th century authors and readers. This might explain the exclusion of occult scientists or at least the focus on scientists who make and discover things in laboratories (see Haynes 1994, on Van Helsing or Sherlock Holmes). At the time of writing *Dracula* (Stoker 1997 [1897]), *In a glass darkly* (Le Fanu 1993 [1872]), or Dr John Silence stories (Blackwood 2002 [1908]), their respective occult detectives were widely recognised as "men of science". There were rather fewer salaried "scientists" in the 19th century than there are today and science was the domain of a variety of professions. Ruth Barton (2003) finds that in the second half of the 19th century the label "men of science" was the term of preference over "scientist" as a phrase that better reflected the multiplicity of routes by which people involved themselves in doing science. So in Barton's terms, socio-cultural constructs of science stepped beyond the confines of a laboratory or a white coat and were based much more upon a process.

19th century culture seems to have identified its scientists in fact and fiction as "men of science" who worked through an investigative process rooted in reason to obtain knowledge for a variety of purposes in a variety of settings. Occult science was as genuine a science as was physics or chemistry. Yet, 21st century scholars generally exclude the occult detective from studies of science in fiction because the laboratory comes transformed perhaps into a cosy fireside location, perhaps the Carpathian mountains or a church vault, and the tool kit is as likely to include a crucifix and a vial of holy water as it is a retort, microscope or thermometer.

That said, if 21st century scholars do not recognise these characters as scientists then is it not likely that readers and consumers do not either? In which case occult detectives are irrelevant except as an oddity of historical science representation. Nonetheless scientists do persist in the supernatural fictions of the 21st century and while they may not be recognised as typified scientists they combine the discourses of science and supernatural. They span a boundary that "enchants science" and "scientises magic", as Simon Locke (2005) finds in his study of comic book superheroes.

Magic and science have a long entwined history that reaches back to past conceptions of nature, spirit and alchemy, and in representation they have remained enmeshed ever since. Mad, bad scientists themselves retain an element of the magical (Locke 2005, 2011; Roach 2011; Schummer 2006; Toumey 1992; Weart 1988) and even the impossibly rational scientific detective Mr Sherlock Holmes can infuse modernity with magic and meaning (Saler 2003). *Frankenstein*, which is after all *the* fiction about science, was originally published by a firm called Lackingtons who specialised in "magic, the illegitimate supernatural, and horror", one advert listing such stock items as: *The Magus or Celestial Intelligencer*; *Lives of the Alchemystical Philosophers*; and *Apparitions, or the Mysteries of Ghosts, Hobgoblins and Haunted Houses* (St Clair 2004, 359).

In addition, Locke (2011) has argued that sociological notions of disenchantment arising from the rationalisation of society of the kind described by Max Weber should not be seen as a uniform state that is simply forced upon the populace. Disenchantment is one facet of a whole kaleidoscope of meanings that science takes on in the social world. Fictional representations comprise one cultural manifestation of such meanings, as are the results of typification, our day to day engagements with technology, or the development of stock characters of science or magic. It seems that wherever discourses of science are present so discourses of the charismatic are never far away, the two apparently in a constant cultural dialogue.[4]

Recent prominent examples of the kinds of "charismatic science" that are employed by occult detectives can be found in a number of texts. Stephenie Meyer's (2005-2008) *Twilight saga* features a vampire family headed by a doctor who polices the vampire world and in a manner is a vampirised version of Professor Van Helsing. *Buffy the vampire slayer* (20th Century Fox Television 1997-2003), Joss Whedon's televised vampire hunter, was assisted by a "Watcher", a kind of parapsychologist who researches the supernatural phenomena that Buffy must confront. James Herbert has given life to a scientific occult detective named David Ash in his novels *Haunted* (2007) and *The ghosts of Sleath* (2010). Even a parodic children's cartoon, *Count Duckula* (Thames Television 1988), retains the Count's scientific nemesis, in this case a non-hubristic mad scientist called Dr Von Goosewing–a comedic Van Helsing.

Positioning Schutz's notion of a typficatory process as phenomenological fabric with which society creates, engages and debates fictions, or indeed knowledge of any kind, makes sense of the seemingly endless reiteration of stock characters and stories of science and supernatural. The fact that focus groups refer to such figures almost as if they were a cultural meta-language suggests that these typified figures do exist as social knowledge and are used as a means of experimenting with the moral, ethical and human implications of science. Such a process also supports Locke's (2011) proposal that the meaning of science in culture is under constant debate. Here then, we have shifted the frame of debate from Adorno's "what does the culture industry do *to* people?" to "what does the culture industry do *for* people?" I do not wish to insist that it has no anaesthetic effect, but I suggest that the very emptiness of stock characters and formula fictions that Adorno so detested is what makes them so useful as ways of debating and experimenting with moral meaning. This is no revolutionary force, but it is a vital socio-cultural resource.

Manufactured Monster and Charismatic Count

Just as the twin demons of supernatural vampire and man-made monster diffuse through the fabric of culture, so they each carry with them the weight of charisma and of science from which they arise respectively. The double discourse is a potent force indeed that gives the duo greater meaning when they are together than either has on its own. Although each story alone does engage both natural and supernatural realms, the polarity of the two stories juxtaposed highlights and magnifies the tension between the two types of force.

Simon Locke (2011, 113) finds that questions of disenchantment or enchantment are a core concern of cultural engagements with science. The continual popular pairing of Frankenstein and Dracula supports his view that the

> combining of science and magic in individual characters and their conflicts is indicative of the tensions surrounding science in modern culture regarding whether it enchants or disenchants and if it is a good or bad thing in either case.

It is also a manifestation of Locke's thesis that by opposing science to magic, the discourse of disenchantment gives life and space to the charismatic or the irrational that it would not have if it were not crucial as science's "other". Neatly put: "the 'irrational' is not eliminated but emerges to greater prominence.... Disenchantment produces and needs enchantment" (Locke 2011, 102).

The potency of the polarity of Dracula and Frankenstein is partly what keeps these figures together then. Devilish charisma is destroyed by science and science creates a man-made monster. Yet despite the tension held in the opposition, at their core each story carries a similar message, which revolves around its orientation to enchantment and disenchantment. Professor Van Helsing points toward the modern splitting of knowledge–science from religion, academic from folk, expert from lay–as being the edge of advantage that the vampire gains in fin de siècle London. In order to lay Count Dracula to rest, the vampire hunter must be both scientist and charismatic. Van Helsing explains:

> We have on our side the power of combination–a power denied the vampire kind; we have the power of science...[and] tradition and superstition.... Does not the belief of vampires rest on them? A year ago which of us would have received such a possibility, in the midst of our scientific, matter-of-fact nineteenth century? (Stoker 1997 [1897], 277)

His point seems to be that science alone does not have the power to fight evil without moorings to a system of knowledge and belief that is older and still enchanted.

In contrast, Victor Frankenstein's lack of superstition is what enables him "to spend days and nights in vaults and charnel houses", examining "every object, the most insupportable to the delicacy of human feelings...how the worm inherited the wonders of the eye and brain"

(Shelley 1994 [1818], 80). Inured as he is to sights that others would find disturbing, he is then able to disregard the feelings of his own creature. Frankenstein's complete split from the realm of feeling and superstition is his tragedy, whereas Van Helsing's character harks back (or forward), almost sentimentally, to a time when all knowledge was one thing and men were wise (and charismatic).

The mad scientist might be seen as arising from anxieties over the splitting of knowledge and if this is so then the vampire hunter must be a direct response to the same anxiety. Joachim Schummer (2006, 125) argues that mad scientists emerge in the 19th century as a result of the loss of a "metaphysical system to provide an overall framework and orientation", which comes about with the gradual fragmentation of knowledge into sciences, arts, humanities, philosophy and perhaps religion too. Van Helsing demonstrates the value of re-joining the circle, while Frankenstein exemplifies the folly of not doing so. They may be mirrors of one another, but their sense is the same and each story reinforces the message of the other–ultimately disenchantment results in and relies upon enchantment. So although the capitalist culture industry may have hitched an opportunist ride on the potency of famous horror twins, the expression of such a weighty dilemma does not have the appearance of a simple anaesthetic for the masses.

The Occult Detectives

The dilemma of disenchantment which opposes the rational and the charismatic defines a point of particularly frenzied cultural activity in fin de siècle England. Not only do period fashions in fiction negotiate the boundaries between the two sources of knowledge through the spread of occult detectives, but there is also a more general drive to test the natural against the supernatural and vice versa by "real world" occult detectives.

Ghost hunts, sometimes drawing thousands of people onto the streets, were a regular occurrence, a likely product of burgeoning newsprint according to Owen Davies (2007). The ghost hunt was very much an urban working class affair, which seemed to shout out the reality of the charismatic, while the courts by this period generally preferred a rationalist disenchanted answer.[5] During the same period, an alternative form of ghost hunt was apparently reserved for the middle classes. Spiritualism was likely seen by many as a more thoughtful and perhaps more decorous route to the spirit world. That said, there is undeniable sensationalism in the act of a medium, who called up spirits that variously talked, rang bells, rapped the table, moved objects, wrote messages and caused people to levitate. Yet spiritualism brought the ghost to the scientist where the supernatural could be subjected to rational enquiry by all manner of different methods (Cooter 1984; Lamont 2004; Luckhurst 2002; Noakes 1999, 2002, 2004; Van Wyhe 2004; Winter 1998).

Both ghost hunts on the street and spiritualist activities behind closed doors involved detective activity of one sort or another, either to find the provenance of the ghost or to investigate the nature of the forces creating the phenomenon on trial. It is not surprising then that novelists invented heroes who could apply rational methods to investigate the supernatural as well. Indeed, the experiment to mix science and supernatural has produced some surprisingly long lasting results particularly in the character of Bram Stoker's vampire hunter and all his cultural spin-offs.

One of the major influences on Stoker's writing is reckoned to be the work of his fellow countryman, J. Sheridan Le Fanu, whose short story collection, *In a glass darkly*, comprises five stories, one of which is his own vampire story, "Carmilla" (Byrne 1973, 92; Skal 1993, 102; Tracy 1993, xxi). They claim their origin from amongst the papers of a deceased, fictional scientist and physician, Dr Martin Hesselius, and are edited and introduced by his secretary who frames each story with its own prologue and epilogue. Le Fanu gives only a brief glimpse of his vampire hunter, Baron Vordenburg, a cameo character compared to Hesselius, who is generally reckoned to be the first example of an occult detective and an inspiration to a whole raft of others including: E. and H. Heron's (aka Hesketh and Katherine Prichard) (2006 [1899]) Flaxman Low; Algernon Blackwood's (2002 [1908]) Dr John Silence; William Hope Hodgson's (2006 [1913]) Thomas Carnacki; and Stoker's Van Helsing as well (Punter 1996).

Compelling character though he is, Dr Hesselius is confined to only the first of Le Fanu's five stories and their linking frames. The first story, "Green tea", sets out as if it were a detective story. Hesselius and reader alike are presented with the enigma of a mild mannered, charitable priest who has an odd secret. Mr Jennings is

> always tolerably well…[but] when he goes down to Warwickshire, to engage in actual duties of his sacred calling, his health soon fails him, and in a very strange way (Le Fanu 1993 [1872], 7).

And so, begins the mystery of the haunted clergyman. Hesselius is at first depicted as every bit the all-knowing, razor sharp detective-cum-doctor. Doctor and vicar meet at a party thrown by Lady Mary Heyduke and, when drawn by the mystery of the vicar, Hesselius scrutinises him with "more minuteness than the ordinary practitioner" finding in him the "likelihood of rewarding inquiry" (Le Fanu 1993 [1872], 8).

Immediately, Jennings is turned "investigated" to Hesselius' "investigator", a factor that makes their relationship manifestly more evocative of detective and suspect than of doctor and patient. The language, too, is couched in terms of suspicion and guilt rather than diagnosis and disease. So, it is "in the agitation of a strange shame and horror" (Le Fanu 1993 [1872], 7) that Jennings sometimes falls ill in the midst of his ministerial duties. There is an element of furtiveness about him, as a man "whose transactions and alarms were carefully concealed with an impenetrable reserve". Yet he is no match for the doctor who "penetrated his thoughts without him being aware of it, and was careful to say nothing which could betray…[his] suspicions" (Le Fanu 1993 [1872], 9). Last, on finding that the priest shares his interest in "metaphysical medicine", Hesselius probes the point and Jennings turns guilty suspect: "He dropped his eyes, and folded his hands together uneasily, and looked oddly, and you would have said, guiltily for a moment" (Le Fanu 1993 [1872], 10).

Armoured with his observations of the "suspect-patient", Hesselius indulges in a display of veritable Sherlock Holmes-style "deductive" logic for Lady Mary, a long-time friend of Jennings. He dazzles her with his inexplicable knowledge. Jennings is a bachelor who has given up writing on a topic of an esoteric nature; he drank a good deal of green tea; and one of his parents once saw a ghost. Lady Mary, confirms him to be correct on all counts, is incredulous and declares the doctor to be quite the "conjurer". He ends this little routine with the affirmation that "I have already conceived a theory about him…I should like much to ascertain from his own lips" (Le Fanu 1993 [1872], 12). So far so good, Dr Hesselius makes an impactful start.

What follows is a long interview between Jennings and Hesselius, in which the debilitated clergyman "confesses" his case history. He is the victim of a haunting that takes the form of a malevolent monkey, which has taken greater and greater hold on his life. Yet, the whole story and circumstance of the haunting is ultimately surplus to Hesselius' requirements because his final assessment of the case is based almost entirely on the key "deductions" he showed off to Lady Mary, with a little Swedenborgian philosophy thrown in.[6] In theory a "metaphysician" is the right person for the job, perhaps more able than most to deal in both physical and spiritual matters. Yet, Hesselius' doctoring function proves to be ineffectual to say the least. Thomas Loe (1998) and Richard Haslam (1998) both make note of the fact that at the very moment when Hesselius is most needed by the beleaguered priest, he is unattainable having taken leave of his usual lodgings so as not to be disturbed whilst deliberating the case.

Yet, Hesselius is widely known as an occult detective (Ascari 2007). Srdjan Smajić (2010) remarks that where he fails as doctor and exorcist, he succeeds as "a detective of the *causes* of occult phenomena and *motivations* for actions and behaviours" (Smajić 2010, 152). Perhaps, Hesselius does have more success in the detective line than in the medical way since he does suggest some potential causal factors. Yet, his over determination of causes–an imbalance of nervous or spiritual fluids, an overdose of green tea, an activation of the "inner eye", nervous congestion complicated by inherited suicidal mania and a lack of confidence in the physician–is a long way from the usual dénouement of the detective story, which demonstrates a single, knowable reality. The detective narrative sets out with many possible realities, which it investigates, weeding out one after another to reveal the one true one at the end.

Unlike "Green tea", Bram Stoker's (1997 [1897]) novel, *Dracula*, does resolve in predictable detective-style fashion. Further, it is narrated in epistolary form, in which letters, diaries and newspaper clippings together tell a story that gives the reader a sense of evidential material and positions her/him as judge and jury. In this, Stoker follows Wilkie Collins' (1999 [1868]) *The moonstone*, which is often cited as the first great detective novel (Farmer 1999; Haycraft 1946). Just as in Collins' novel, so in Stoker's the story passes from one witness to another and a mystery unfolds that impels detection. Stoker's narrative in fact has a more insistent lead detective, Van Helsing, assisted by his sidekick, the alienist, Dr Seward. Originally the Professor was conceived not as one, but as three individuals, a detective inspector and two historians (Frayling 1991, 305). In the event, Van Helsing is not a detective by profession, although his mission entails tracking a known deviant by logical, empirical means in much the same way that any criminal investigator might do.

Nearly all detective narratives take a similar progression. The story begins with a crime and sets out on the drudgery of fact collecting with a gradual elimination of suspects until momentum and excitement quicken nearing resolution, when the facts all come together to reveal the perpetrator and history of the crime. Joseph Agassi (1982, 101) has noticed that this is the approach recommended by Francis Bacon as the way to achieve solid scientific progress. It amounts to induction, the

> patient gathering of masses of details and their repeated examination until, perhaps with the aid of a systematic elimination of false possibilities, facts

> fall in to a pattern by themselves, until the pattern or law of nature clicks, as it were.

In Bacon's view a guiding hypothesis or preconceived idea ran the risk of obscuring potentially significant data points and valuable clues by suggesting shortcuts in fact gathering. Of course, fictional detectives themselves do not always adhere to these rules, but the arc of their story almost always matches a narrative of Baconian induction. Fundamental to such a narrative, both in inductive science and in literature is the move from a disordered world (of facts or society), to an ordered world (of understanding or retribution). The Baconian narrative restores order and might be described as a restorative narrative.

"Green tea", however, does not fit the same pattern. There is little or no data gathering to speak of. Hesselius appears to make his final inferences at the outset whilst chatting to Jennings "upon indifferent subjects" at a cocktail party. This is apparently the sum of Hesselius' enquiry and the intervening long interview with Jennings does not seem to furnish him with further details that allow him to either eliminate false possibilities or to find new ones. Loe (1998, 301) suggests that in juxtaposing the

> lack of causality in the sequence of events to Hesselius' supposedly rational inquiry, one can see how Le Fanu strengthens the effect of the story by having one type of structure countering the other.

The result opposes a Baconian narrative and produces what Loe calls a "strange modernism" in which authority is raised only to be undermined. The story begins with a simple question: What is wrong with Mr Jennings? It ends with the same question plus a whole array of potential answers and offers no way of choosing between them. In this regard, "Green tea" ends with a more disordered world than the one it started with. A question that seemed simple at the outset turns into a labyrinthine puzzle by the end.

"Green tea" and the following two of Le Fanu's stories are, in fact, rare examples of Tzvetan Todorov's (1975) genre, "the fantastic" in which supernatural events are strongly indicated but the story allows the possibility that events may have natural provenance. This means that tales of "the fantastic" must contain dual narratives from which the reader chooses: "one probable and supernatural, the other improbable and rational" (Todorov 1975, 49). Is Mr Jennings haunted or is he mentally ill? These are the alternative stories. Todorov's "fantastic" is actually the moment of hesitation, or wavering, between two such possibilities–haunted or mentally ill?

A hero of the hesitant ground of the "fantastic", as Hesselius is, cannot be an effective detective because the fantastic deals only in possibilities that are left hanging, unresolved into a single, identifiable reality. Although Stoker took some of his cues from Sheridan Le Fanu, his own story is much simpler. The dreaded Count Dracula is unequivocally evil and supernatural, a deviant demon whose final retribution has him crumbling to dust. Stoker's Professor Van Helsing demonstrates the merits (and proper use) of a rational method, whereas Le Fanu's Hesselius, though billed as a man of "immense knowledge" and a healer, fails to heal while he succeeds in identifying several possible realities that are never resolved into one. If anything, Le Fanu approaches the depths of complexity and conflict present in life and questions the nature of reality in doing so. The "real" world is disordered and realities are multiple. On this count, Theodor Adorno may have had some sympathy for Le Fanu.

Yet Le Fanu's text is not so easily bent and moulded to provide the substance for the stock of social knowledge. Neither is it easily employed as a framing device; because of its complexity any attempt to frame debate with reference to Hesselius leads back to questions about Hesselius' story itself. The simplicity of the shape and ideas in *Dracula*, however, function rather better in this setting. They are easily taken out of context and projected into other situations–the vampire might be taken as a disease and the debate becomes does the doctor take folk remedies seriously? Does the doctor need the "power of combination" to fight disease? Much in the same way, the mad scientist can raise questions about the possibility of being over-technical.

It seems likely then that fictional images perform different functions depending upon on their own "density". This is a term used by literary critic, Tzvetan Todorov (1975) who warns that the idea of representation itself needs to be re-examined with respect to literary texts, which are not representative in the sense of referring to "a thing" in the world outside the text, in the way that a "factual" statement refers to "something" exterior to the statement itself. The events and characters of literature are only internal to the text, they both exist and they do not exist. Todorov has usefully elaborated on this difficulty and is of the view that because fiction describes events and characters that have parallels in the real world, to deny it any representative role at all would amount to defining an object only with the object itself. There must be at least an oblique relationship between real world and literary representations.

Unlike a word, the fictional image itself is not a signifier, "because it signifies itself…it possesses a certain density" and if we negate such density, then every image turns into allegory, which overstates its referential role and treats it as if it were transparent (Todorov 1975, 143-144). The sense that arises from a comparison of "Green tea" and *Dracula* is that some fictional images have a greater transparency than others, which gives them wider function as socio-cultural resources outside of their own literary text–and such is a function of the "culture industry".

Narratives of Nature and Supernature

The application of scientific method to the supernatural realm both in practice and in literature is sometimes seen as indicative of a "shoring up" of scientific authority (Ascari 2007). Oddly enough, the project largely had the opposite effect and occult detectives were somewhat compromised in rational, scientific terms, while the charismatic and enchanted gained credibility from a gloss of rationalisation. Supernatural investigators often need to rely on all those skills formally disassociated from scientific method, such as subjectivity, their feeling nature and perhaps the power of their own souls. These characters were very popular in fiction for a short period spanning the turn of the 19th and 20th centuries, but have not been so widely and regularly re-worked as has the criminal detective or the mad scientist.[7] Yet the experiment of placing detectives in the occult did create a mix that ultimately allowed science to piggyback the vampire story, one of the most popular genres ever to have existed in fiction.

Aside from Le Fanu's fantastic, the fin de siècle occult detectives generally follow the narrative arc of Bacon's induction. For example, Algernon Blackwood's (2002 [1908]) Dr John Silence begins his first investigation much as Sherlock Holmes would. In "Case 1: a psychical invasion", Silence is presented with the problem, a haunting, he then proceeds by gathering data about the house and the habits of the current inhabitants and he gradually builds a picture of the case. He diagnoses psychical invasion owing to an overdose of cannabis and brings the case to a denouement by using his own psychic methods to draw out the evil and dissipate it.

Just as the fantastic has a dual narrative, so does the detective story as Todorov (1977) has pointed out in a now famous essay entitled "The typology of detective fiction". Here, he (1977, 44) stipulates that the detective story contains "not one but two stories: the story of the crime and the story of the investigation". Todorov (1977, 45) explains that we may

> characterise these two stories by saying that the first–the story of the crime–tells 'what really happened', whereas the second–the story of the investigation–explains 'how the reader (or the narrator) has come to know about it'.

In other words, the first story consists of events as they happened and the second the investigation.

Furthermore, these two stories correspond to ways in which readers make sense of narratives in general. These were initially described by Russian formalists and relate to the way that readers deal with distortions of time, such as flashbacks, flash-forwards, or perhaps even sideways moves (omniscience).[8] At a basic level, narrative structure consists of the order of events in which they are given by the text–the sjužet–which includes all its anachronies of time-shift and is sometimes referred to as plot. Then there is the chronological sequence of events, which in most cases differs from the order given in the sjužet and is constructed by the reader–this is the fabula. What Todorov notices is that, in detective fiction, the story of the crime, or "what really happened" corresponds to fabula, and the detective's investigation corresponds to sjužet, the story as it is written. Put another way, the detective derives the "truth" of what really happened from the clues s/he finds just as the reader derives the fabula from the sjužet.

In detective fiction the two narrative structures work together, the one supporting the other, so that the Baconian shape, the fact gathering and elimination of false possibilities, gradually positions elements of fabula with increasing impetus in their chronological order. This effectively situates the detective between the reader (and sjužet) and the fabula–in a sense, the detective becomes "guardian" of the fabula. Seen in this light, the notion of restorative narrative takes on new meaning. It is only through the detective that the reader is able to reconstruct or to re-order the narrative to fit its "true" chronological pattern. Likewise, the scientist constructs a fabula from Nature's sjužet.

Indeed the advent of detective fiction in the 19th century is congruent with the rise in historical sciences. Charles Rzepka (2005) points to one of the most influential of popular scientific developments of the 19th century, Charles Lyell's explication of uniformitarianism in *Principles of geology*, published in 1830, as a key development in the making of a detective narrative. In this, Lyell argues that the earth's structure can, or should, only be understood by recourse to processes visible in the present that had acted in the past with the same degree of intensity. Such cause and effect thinking, popularised in *Principles*, may have a cue for the detective narrative, because it posits an orderly nature *and* an orderly process of nature, which meant that the present is the sure key to the past.

Later in the century, the tables are turned when T.H. Huxley (1896) describes scientific methods in his essay "On the method of Zadig", by drawing on Voltaire's detective, Zadig, from the 1747 story of the same name, to exemplify the process of science. The observations of scientists, Huxley (1896, 106-8) says, are "retrospectively prophetic and strive towards the reconstruction in human imagination of events which have vanished and ceased to be". Perhaps even more powerfully, he (1896, 131-3) continues

> inasmuch as…retrospective prophecies are the result of following backwards, the very same method as that which invariably leads to verified results, when it is worked forwards…[they provide] as much reason for placing full confidence in the one as in the other.

Science, then is the method by which the past can be (re)constructed and the future too, a means for establishing a narrative of nature. Arthur Conan Doyle (2008 [1892], 114) places the detective method on the same plain, when Sherlock Holmes pontificates

> [t]he ideal reasoner…would, when he had once been shown a single fact in all its bearings, deduce from it not only all the chain of events which led up to it but also all the results which would follow from it.

The fixed order of time then is a vital prerequisite for such a rational method as this. Chris Morash (1998, 123) has pointed out a key irrationality of the supernatural as "a grotesque, disruption of the linearity of time and hence of reason itself". A ghost, after all, represents the dead returned. This particular feature of the supernatural makes for a distinct narrative structure. Using the example of Charles Dickens' *A Christmas carol*, Morash (1998, 131-132) demonstrates that movements in time in the supernatural narrative, between Christmases past, present and future "are not presented in such a way that we as readers are asked to put them back in their correct chronological sequence when constructing a fabula". Scrooge's travels in time are happening in the present of the story. Similarly, the ghosts of past, present and future are incursions from different times into the present. For, Morash, this is a key feature of the supernatural tale. What it amounts to is that time is allowably disordered in the fabula, something that does not happen in non-supernatural fiction, which only permits anachrony to exist at the level of sjužet.

According to Sherlock Holmes, the detective is tasked with ironing out the anachrony of the sjužet, so that detective first, then reader can construct a fabula by "placing upon record that severe reasoning from cause to effect" (Conan Doyle 2008 [1892], 270). In a supernatural world, fabula and sjužet are collapsed into one and so the rational, scientific detective is effectively redundant. Huxley (1896, 135), too, reminds us that the establishment of cause and effect relies upon "[t]he *constancy* of the *order* of nature, being the common foundation of all scientific thought". Here then is a difficulty for the occult detective. The investigation of a realm, the supernatural, which does not have the same "constancy" of "order" makes it impossible to reason from cause to effect and vice versa.

Such problems are well demonstrated in a case study by Susan Hoyle (2004) on the declining credibility of the witchcraft narrative in 19th century England. She argues that it was not so much that the claims of science stole a march on the credibility of magic, or even gained greater respect, rather it was a falling off in acceptability of the kind of story that the witchcraft narrative was prone to tell. At the same time, she (2004, 46) finds a "rise in acceptability of a forensic narrative based on the demonstration of detective skills". She illustrates her case with evidential statements reported from court proceedings and, though her empirical evidence is sparse, she presents two cases that show how natural and supernatural narratives talk past one another. She also suggests why and under what circumstances one narrative is preferred over another.

One case involved a charge of wounding–a witch-cutting. The Davis family believed themselves to have been cursed by a neighbour and since the traditional means of breaking a spell is to draw blood from the witch, John Davis cut his neighbour's cheek. According to the Davis's, the curses had taken the form of repeated instances of haunting. The case seemed to hang on one of these, in which an invalid daughter was assaulted when, as reported in *The Times*, November 4, 1867,

> a man and woman came down the chimney…both persons headless, and seized her by the body, cast her violently on the ground, and then tossed her in the air, after which they took the sofa she had used and went through a similar feat with it.

This was the story given in John Davis' defence. The prosecution case rested upon the evidence of "Police-Superintendent Richardson [who] showed that the accumulated dust around the legs of the furniture proved that no such thing could have happened". As Hoyle points out, the two narratives talk past one another. Firstly, the fact that the dust was not disturbed was no evidence against the presence of headless spirits since the magic that called up the spirits was surely able to enchant the dust. Anyway, the dust narrative only works under natural material conditions when the order of time is constant.

Similarly, regarding "Green tea", Loe (1998, 301) juxtaposes the "lack of causality in the sequence of events to Hesselius' supposedly rational inquiry". Indeed to combine a narrative that at heart is reliant on the linearity of time with another that, by definition, distorts the time dimension seems almost guaranteed to create the effect of the fantastic, since the reader has a choice of narratives each talking past the other. Yet many of the scientific supernatural investigators who followed Le Fanu's example were not, as he was, aiming to create such an effect. These authors were more likely swept along by the increased popularity and credence of a narrative that demonstrated detective skills and harmonised with the propensity of readers to order time in re-order as Hoyle (2004) has illustrated within the legal system.

In the case of the witch-cutting, the court allowed the evidence about the dust and the papers reported it as legitimate evidence against the haunting even though it was no proof against witchcraft which was not on trial anyway (the cutting was). To demonstrate the opposite effect, Hoyle goes on to report a second case in which a known "cunning man" accused of poisoning was not convicted despite the fact that there was ample evidence against him.[9] In this case, the forensic evidence was poorly marshalled by an inexperienced surgeon who contradicted himself. Hoyle (2004, 59) argues that, "the prosecution for drugging failed because it did not have the services of someone who could tell a good forensic story". What these two cases reveal is that when a forensic, or cause and effect, narrative was adequately marshalled, it was more persuasive than a witchcraft narrative in the courts at the time (so it has likely remained today).

That the belief in the supernatural did not wane in the face of the spread of scientific ideas is, apart from anything else, clear from the spiritualist revival and plethora of ghost clubs that followed from it. Science itself is unlikely to have caused a decline, observable through court reports, in witchcraft belief across the 19th century. Hoyle suggests that what changes people's belief is at least partly to do with fashions in storytelling. The story that came to be preferred in the courts and elsewhere as evidenced by the spread of detective fiction was the story of process, or of method and its concomitant demonstration of cause and effect that conformed to the reader's desire to place events in a chronological order.

Perhaps as a result of the growing fashion for cause and effect stories, several occult detectives and scientists followed from Le Fanu's example, a factor that has sometimes been interpreted as the triumphant note for science and secularisation (Ascari 2007). Yet this is unlikely and the frenzy of activity at the fin de siècle that pitted reason against unreason and rational against charismatic is simply another episode in the dance of disenchantment. In any case, it is more likely, given Hoyle's analysis of storytelling fashions, that the aim was not to proclaim the authority of science over supernatural, but rather to give the supernatural a fashionable gloss of scientific verisimilitude–one that would convince "our scientific, matter-of-fact nineteenth century".

Ultimately, cause and effect reasoning is destabilised in the face of the supernatural. So that in such experiments as Algernon Blackwood's Dr John Silence, and E. and H. Heron's Flaxman Low stories, detectives are forced into using other means than reason, no matter how insistent they are on the natural provenance of events at hand. For example, Dr John Silence could "vicariously absorb these evil radiations into himself and change them magically into his own good purposes" (Blackwood 2002 [1908], 824). And occult detective Flaxman Low, who insists there is only natural causation, says of one case

> everybody who, in a rational and honest manner, investigates the phenomena of spiritism will, sooner or later, meet in them some perplexing element, which is not to be explained by any of the ordinary theories (Heron and Heron 2006 [1899], 336).

That said, the occult detective is the vampire hunter, a figure that has spawned a variety of modern ghostbusters who pit science against supernatural to varying degrees. Of all supernatural investigators it appears to be the vampire hunter who has grown to mythic proportion.[10] The trailblazer for this breed of investigator is Dracula's destroyer. The advance that Stoker makes on other psychic investigators which gives his novel an intuitive coherence that the others do not have is that his ghoul, though through and through supernatural, has a material body. The Count's materiality allows the team to combine vampire lore with natural law to create more than a semblance of a cause and effect procedural investigation. What is fascinating about the success of this story is that it manages to retain the detective process intact because the detectives or the vampire hunters are able to research and track a material entity.

Once on their quarry vampire hunters use a kind of magic that has elements of a science anyway, a "scientised" magic (Locke 2005, 33). Vampire lore has clear-cut methodological rules that can be rote learned and do not rely upon the subjectivity or soul of the detective or hunter. What emerges from Stoker's handling of science and supernatural is a structure and an outcome that proclaims the power of belief while it demonstrates the power of science (Jann 1989). Given the difficulty of bringing together two worldviews that talk past one another, this was and still is an unexpected gift for popular science and allows us, with its partnered text, *Frankenstein*, to continue to debate the dilemmas of disenchantment.

Notes

1. "mad scientist *n.* a scientist who is mad eccentric, esp. so as to be dangerous or evil: a stock figure of melodramatic horror stories; freq. *attrib.*" *Oxford English dictionary*.
2. In his introduction to the new English version of *Grimm's fairy tales*, Philip Pullman (2012, xiii) writes that "conventional stock figures" in these stories "are not actually conscious". In other words they are barely human. Though not as extreme, the mad scientist tends towards being similarly hollow as do Count Dracula and Professor Van Helsing.
3. Such images are well documented from so-called "draw-a-scientist" studies, for example, Rahm and Charbonneau (1997), Wade-Chambers (1983), and in Mead and Métraux's (1957) survey. That said, the disaster narrative is by no means fused to this image and some scientists make positive use of the image. See, for example, Professor Martin Poliakoff, presenter of short chemistry videos at http://www.periodicvideos.com/about.htm and the education group at http://www.madscience.org/.
4. I use the term "charismatic" after Locke (2011). It refers to Weber's use of the term "extraordinary power" from which he derives a belief in spirits as described by Whitehead (1974, 552).
5. A rumoured haunting at No. 6 Pond Place in Chelsea drew crowds such that, as reported in *The Times*, September 12, 1853, "the neighbourhood had been in a state of uproar…it had been necessary to send a number of the police to disperse the crowds". A follow up article in the same newspaper reports a séance in which the spirit of "Shelley, the poet" confirmed that there was a spirit in No. 6 Pond Place. *The Times*, May 27, 1865, reported large crowds causing nightly disturbances at St George's Church, Southwark and on June 1, 1867, the same happening at Woburn Square, London. Crowds of between 5,000-6,000 people were reported in *The Times*, July 6, 1874, to have converged nightly on Christ Church, Westminster until a PC Thew apprehended a man running across the graveyard with a sheet over his head. *The Times*, July 6, 1876 reported crowds of 300-400 people congregating in New Weston Street, Southwark hoping to see the ghosts of notorious murderers the Mannings who had lived nearby.
6. Hesselius is a medic with an interest in "metaphysical medicine" and in this respect owes a debt to the Swedish scientist cum Christian mystic, Emanuel

Swedenborg (1688-1772). Fundamental to Swedenborg was the idea that life exists on dual planes of existence, the spiritual and the physical.

7. Of course, there are exceptions to the rule, for example James Herbert's detective David Ash from his novels *Haunted* (2007) and *The ghosts of Sleath* (2010). Ash is a quintessential "hard boiled" detective as a somewhat damaged, hard-drinking loner and originally a sceptic until he realised his own psychic powers.

8. Todorov (1977) uses the terms "fable (story)" and "subject (plot)", but it is often conventional to use the original formalist terms, fabula and sjužet, as does Morash (1998).

9. Cunning folk were one of the magical trades, who variously sold love-potions, found lost items, recovered stolen property and warded off witches. See Davies (2007).

10. There are other types of narrative, for example, *Ghostbusters* (1984), a comedic feature film about parapsychologists, which turns the 'ghostbuster' into a joke of sorts. A BBC television drama series, *Apparitions* (2008) and William Peter Blatty's *The exorcist* (1971) return the supernatural to the professional expertise of the clergy rather than scientists. However, many more popular supernatural investigators are based on the investigative vampire hunter, including: Joss Whedon's *Buffy the vampire slayer* (1997-2003), who destroys a whole range of ghouls; the eponymous hero of the film *Van Helsing* (2004); Stephenie Meyer's (2005-2008) *Twilight saga*, about a vampire family who police the vampire world and in a manner are vampirised versions of Van Helsing; and James Herbert's *Haunted* (2007) and *The ghosts of Sleath* (2010), featuring an occult detective.

References

Adorno, Theodor W. 1975. Culture industry reconsidered. Trans. Anson G. Rabinbach. *New German Critique* 6: 12-19.

Adorno, Theodor W., and Max Horkheimer. 1997 [1944]. The culture industry: enlightenment as mass deception. In *Dialectic of enlightenment*. Trans. J. Cumming, 120-167. London: Verso.

Agassi, Joseph. 1982. The detective novel and scientific method. *Poetics Today* 3: 99-108.

Ascari, Maurizio. 2007. *A counter history of crime fiction*. Basingstoke: Palgrave Macmillan.

Blatty, William Peter. 1971. *The exorcist*. New York: Harper and Row.

Baldick, Chris. 1987. *In Frankenstein's shadow: myth, monstrosity and nineteenth century writing*. Oxford: Clarendon Press.

Barton, Ruth. 2003. "Men of science": language, identity and professionalization in the mid-Victorian scientific community. *History of Science* xli: 73-119.

Basalla, George. 1976. Pop science: the depiction of science in popular culture. In *Science and its public: the changing relationship*, eds.

George Holton and William A. Blanpied, 261-278. Dordrecht: D. Reidel.
Bates, Benjamin R. 2005. Public culture and public understanding of genetics: a focus group study. *Public Understanding of Science* 14: 47-65.
Blackwood, Algernon. 2002 [1908]. *John Silence, psychic detective*. Berkeley CA: Renaissance E Books. Kindle edition.
Byrne, Patrick F. 1973. Joseph Sheridan Le Fanu: a centenary memoir. *Dublin Historical Record* 26: 80-93.
Cohen, I. Bernard. 1981. The fear and distrust of science in historical perspective. *Science, Technology, & Human Values* 6: 20-24.
Collins, Wilkie. 1999 [1868]. *The moonstone*, ed. Steve Farmer. Peterborough, CA: Broadview Press.
Conan Doyle, Arthur. 2008 [1892]. *The adventures of Sherlock Holmes*, ed. Richard Lancelyn Green. Oxford: Oxford University Press.
Cooter, Roger. 1984. *The cultural meaning of popular science: phrenology and the organisation of consent in nineteenth-century Britain*. Cambridge: Cambridge University Press.
Dalziel, Margaret. 1957. *Popular fiction 100 years ago: an unexplored tract of literary history*. London: Cohen and West.
Davies, Owen. 2007. *The haunted: a social history of ghosts*. Basingstoke: Palgrave Macmillan.
Emsley, John. 2001. Good news is no news. *Nature* 413: 113.
Farmer, Steve. 1999. Introduction to *The moonstone*, by Wilkie Collins, 9-34. Peterborough, CA: Broadview Press.
Frayling, Christopher. 1991. *Vampyres: Lord Byron to Count Dracula*. London: Faber and Faber.
—. 2005. *Mad, bad and dangerous?: The scientist and the cinema*. London: Reaktion Books.
—. 2006. All boffins are bonkers. *Telegraph*, May 9. http://www.telegraph.co.uk/connected/main.jhtml?xml=/connected/2006/05/09/ecboff09.xml&sSheet=/connected/2006/05/09/ixconnrite.html
Gilbey, John. 2008. Fools! I will destroy you all! *Times Higher Education Supplement*, June 19. http://www.timeshighereducation.co.uk/story.asp?storycode=402421
Haslam, Richard. 1998. Le Fanu and the fantastic semantics of ghost-colonial Ireland. In *That other world: the supernatural and the fantastic in Irish literature and its contexts Volume 1*, ed. Bruce Stewart, 267-284. Gerards Cross: Colin Smythe.

Haste, Helen. 1997. Myths, monsters, and morality: understanding 'antiscience' and the media message. *Interdisciplinary Science Reviews* 22: 114-120.

Haycraft, Howard, ed. 1946. *The art of the mystery story: a collection of critical essays*. New York: Carroll and Graf.

Haynes, Roslynn. 1994. *From Faust to Strangelove: representations of the scientist in western literature*. Baltimore: Johns Hopkins University Press.

—. 2003. From alchemy to artificial intelligence: stereotypes of the scientist in western literature. *Public Understanding of Science* 12: 243-253.

—. 2006. The alchemist in fiction: the master narrative. *HYLE: International Journal for Philosophy of Chemistry* 12: 5-29.

Herbert, James. 2007. *Haunted*. London: Pan-Macmillan.

Herbert, James. 2010. *The ghosts of Sleath*. London: Harper Collins. Kindle edition.

Heron, E., and H. Heron. 2006 [1899]. *Flaxman Low: occult psychologist*. Kindle edition. http://gutenberg.net.au/.

Hodgson, William Hope. 2006 [1913]. *The casebook of Carnacki the ghost finder*, ed. David Stuart Davies. Ware: Wordsworth.

Hoyle, Susan. 2004. The witch and the detective: mid-Victorian stories and beliefs. In *Witchcraft continued: popular magic in modern Europe*, eds. Willem de Blécourt and Owen Davies, 46-68. Manchester, Manchester University Press.

Huxley, Thomas Henry. 1896. *On the method of Zadig*. Kindle edition. http://www.gutenberg.org/.

Jackson, J. Kasi. 2008. Gender, mad scientists and nanotechnology. *Spontaneous Generations* 2: 45-55.

Jann, Rosemary. 1989. Saved by science? The mixed messages of Stoker's *Dracula*. *Texas Studies in Literature and Language* 31: 273-287.

Jones, Robert. 2001. "Why can't you scientists leave things alone?" Science questioned in British films of the post-war period (1945-1970). *Public Understanding of Science* 10: 365-382.

Jones, Stephen. 1993. *The illustrated vampire movie guide*. London: Titan Books.

Jörg, Daniele. 2003. The good, the bad and the ugly–Dr. Moreau goes to Hollywood. *Public Understanding of Science* 12: 297-305.

Kirby, David. 2003. Scientists on the set: science consultants and the communication of science in visual fiction. *Public Understanding of Science* 12: 261-278.

Kramer, Cheryce. 2003. Introduction. In *Literature and science 1660-1834: volume I, science as polite culture*, eds. Cheryce Kramer, Trea Martyn and Michael Newton, xi-xxxviii. London: Pickering and Chatto.

Lamont, Peter. 2004. Spiritualism and a mid-Victorian crisis of evidence. *The Historical Journal* 47(4): 897-920.

Le Fanu, Sheridan J. 1993 [1872]. *In a glass darkly*, ed. Robert Tracy. Oxford: Oxford University Press.

Locke, Simon. 2005. Fantastically reasonable: ambivalence in the representation of science and technology in super-hero comics. *Public Understanding of Science* 14: 25-46.

—. 2011. *Re-crafting rationalization: enchanted science and mundane mysteries*. Farnham: Ashgate.

Loe, Thomas. 1998. The strange modernism of Le Fanu's "Green tea". In *That other world: the supernatural and the fantastic in Irish literature and its contexts volume 1*, ed. Bruce Stewart, 293-306. Gerards Cross: Colin Smythe.

Luckhurst, Roger. 2002. *The invention of telepathy*. Oxford: Oxford University Press.

McDonald, Kim. 1989. A Stanford President warns researchers of the need to combat the American public's "alarming" mistrust of science. *The Chronicle of Higher Education*, January 25: A4.

Mead, Margaret, and Rhoda Métraux. 1957. Image of the scientist among high-school students: a pilot study. *Science* 126: 384-390.

Meyer, Stephenie. 2005-2008. *Twilight saga*. New York: Little Brown and Co.

Michael, Mike, and Simon Carter. 2001. The facts about fictions and vice versa: public understanding of genetics. *Science as Culture* 10: 5-32.

Millhauser, Milton. 1973. Dr Newton and Mr Hyde: scientists in fiction from Swift to Stevenson. *Nineteenth-Century Fiction* 28: 287-304.

Morash, Chris. 1998. The time is out of joint (o cursèd spite). In *That other world: the supernatural and the fantastic in Irish literature and its contexts Volume 1*, ed. Bruce Stewart, 123-142. Gerards Cross: Colin Smythe.

Mulkay, Michael. 1995. Parliamentary ambivalence in relation to embryo research. *Social Studies of Science* 25: 149-163.

—. 1996. Frankenstein and the debate over embryo research. *Science, Technology, & Human Values* 21: 157-176.

Noakes, Robert J. 1999. Telegraphy as an occult art: Cromwell Fleetwood Varley and the diffusion of electricity to the other world. *British Journal for the History of Science* 32: 421-459.

—. 2002. Instruments to lay hold of spirits: technologizing the bodies of Victorian spiritualism. In *Bodies/Machines*, ed. Iwan Rhys Morus, 125-164. Oxford: Berg.

—. 2004. Spiritualism, science and the supernatural in mid-Victorian Britain. In *The Victorian supernatural*, eds. Nicola Brown, Carolyn Burdett and Pamela Thurschwell, 23-43. Cambridge: Cambridge University Press.

Orthia, Lindy A. 2010. Anti-rationalist critique or fifth column of scientism? Challenges from Doctor Who to the mad scientist trope. *Public Understanding of Science OnlineFirst* 1?: 1-18. doi: 10.1177/0963662509355899.

Pollack, Andrew. 1998. Scientists seek a new movie role: hero, not villain. *New York Times*, December 1.

Pullman, Philip. 2012. *Grimm tales for young and old*. London: Penguin.

Punter, David. 1996. *The literature of terror: the Gothic tradition Volume 1*. Harlow: Pearson Education.

Rahm, Jrène, and Paul Charbonneau. 1997. Probing stereotypes through students' drawings of scientists. *American Journal of Physics* 65: 774-778.

Roach, Katherine. 2011. Between magic and reason: science in 19th century popular fiction. PhD diss., University of Nottingham.

Rohn, Jenny. 2006. Experimental fiction. *Nature* 439: 269.

Rose, Christopher. 2003. How to teach biology using the movie science of cloning people, resurrecting the dead, and combining flies and humans. *Public Understanding of Science* 12: 289-296.

Rzepka, Charles J. 2005. *Detective fiction*. Cambridge: Polity Press.

Saler, Michael. 2003. "Clap if you believe in Sherlock Holmes": mass culture and the re-enchantment of modernity, c.1890-c.1940. *The Historical Journal* 46: 599-622.

Schummer, Joachim. 2006. Historical roots of the "mad scientist": chemists in nineteenth-century literature. *AMBIX* 53: 99-127.

Schutz, Alfred. 1967. *The phenomenology of the social world*. Trans. G. Walsh and F. Lehnert. Evanston: Northwestern University Press.

—. 1973. *Collected papers I: the problem of social reality*, ed. Maurice Natanson. The Hague: Martinus Nijhoff.

Shelley, Mary. 1994 [1818]. *Frankenstein*, eds. D.L. Macdonald and Kathleen Scherf. Peterborough, CA: Broadview Press.

Shiach, Morag. 1989. *Discourse on popular culture: class, gender, history in cultural analysis, 1730 to the present*. Cambridge: Polity Press.

Skal, David J. 1993. *The monster show: a cultural history of horror*. London: Plexus.

—. 1998. *Screams of reason: mad science and modern culture*. New York: Norton and Co.

Smajić, Srdjan. 2010. *Ghost-seers, detectives, and spiritualists*. Cambridge: Cambridge University Press.

St Clair, William. 2004. *The reading nation in the romantic period*. Cambridge: Cambridge University Press.

Stoker, Bram. 1997 [1897]. *Dracula*, ed. Glennis Byron. Peterborough, CA: Broadview Literary Texts.

Todorov, Tzvetan. 1975. *The fantastic: a structural approach to a literary genre*. Trans. Richard Howard and Robert Scholes, Robert. Ithaca: Cornell University Press.

—. 1977. *The poetics of prose*. Oxford: Basil Blackwell.

Toumey, Christopher P. 1992. The moral character of mad scientists: a cultural critique of science. *Science, Technology, & Human Values* 17(4): 411-437.

Tracy, Robert. 1993. Introduction to *In a glass darkly*, by Sheridan Le Fanu, vii-xxviii. Oxford: Oxford University Press.

Tudor, Andrew. 1989. Seeing the worst side of science. *Nature* 340: 589-592.

Turney, Jon. 1998. *Frankenstein's footsteps: science, genetics and popular culture*. New Haven: Yale University Press.

Van Wyhe, John. 2004. *Phrenology and the origins of Victorian scientific naturalism*. Aldershot: Ashgate.

Wade-Chambers, David. 1983. Stereotypical images of the scientist: the draw a scientist test. *Science Education*, January: 255-265.

Weart, Spencer. 1988. The physicist as mad scientist. *Bulletin of Science, Technology and Society* 8: 143-152.

Weingart, Peter. 2006. Chemists and their craft in fiction film. *HYLE: International Journal for Philosophy of Chemistry* 12: 31-44.

Weingart, Peter, Claudia Muhl and Petra Pansegrau. 2003. Of power maniacs and unethical geniuses: science and scientists in fiction film. *Public Understanding of Science* 12: 279-287.

Weingart, Peter, and Petra Pansegrau. 2003. Introduction: perception and representation of science in literature and fiction film. *Public Understanding of Science* 12: 227-228.

Whitehead, Harriet. 1974. Reasonably fantastic: some perspectives on Scientology, science fiction and occultism. In *Religious movements in contemporary America*, eds. I.I. Zaretsky and M.P. Leone, 547-587. Princeton, NJ: Princeton University Press.

Winter, Alison. 1998. *Mesmerized: powers of mind in Victorian Britain*. Chicago: Chicago University Press.

CONTRIBUTORS

Alice Bell is a research fellow and head of public engagement at the Science Policy Research Unit, University of Sussex, focusing on public engagement with science and technology. She is a co-editor at New Left Project and also co-runs the Brain Train podcast. Alice has published on children's science books; anachronism and time travel in contemporary culture; and communities of science bloggers. She also writes about science in society issues for the *Guardian*, *Popular Science* and the *Times Higher Education*.

John Besley is an associate professor and the Ellis N. Brandt Chair in Public Relations within the College of Communication Arts and Sciences at Michigan State University. His research explores the relationships between media use, public engagement and health/environmental risk perceptions and public perceptions of nanotechnology, biotechnology and energy technologies.

Meg Blanchard is Associate Professor of Science Education in the STEM Education Department at North Carolina State University, and Research Director of The Science House at NC State in Raleigh, NC. Her current research investigates the effect of a systemic reform model on career awareness, interest and preparation. Meg has conducted teacher professional development and research in FL and NC with grants from the National Science Foundation, the U.S. Department of Education and the BelleJAR Foundation. She was recently inducted into the university's Outstanding Extension & Service Academy and received the 2013 Alumni Association Outstanding Extension and Outreach Award for NC State. Meg has published in *Science Education*, the *Journal of Research in Science Teaching*, the *International Journal of Science Education*, *Research in Science Education* and also in practitioner journals of the National Science Teacher Association.

Mark B. Brown is Associate Professor in the Department of Government at California State University, Sacramento. He was a postdoctoral fellow at the Institute for Science and Technology Studies, Bielefeld University. Mark is the author of *Science in Democracy:*

Expertise, Institutions, and Representation (MIT Press, 2009) and has written various book chapters and journal articles on the politics of expertise, citizen participation, bioethics, climate change and related topics.

Miquel Carandell is currently undertaking a PhD at the CEHIC exploring the history of two paleoanthropological and archaeological controversies in contemporary Spain and Brazil, and is also working in the science popularisation online magazine *UAB Divulga*. He has been a visiting researcher at the Max Planck Institute for the History of Science and is a member of the PhD students group in Barcelona (XeHC!), the Catalan Society seminar series organising committee, and the Science and Technology in the European Periphery (STEP) Steering Committee.

Michael Cobb is Associate Professor of Political Science in the School of Public and International Affairs at North Carolina State University. His research spans a diverse set of topics from the perspective of a political psychologist, including public opinion about emerging technologies, trade and war. He is currently studying misinformation in public opinion and public acceptance of using genetically engineered mosquitoes to control the spread of dengue fever. His most recent publications are in the journals of *Political Psycology*, *Nanoparticles Research* and *Conflict Resolution*.

Elaine Howard Ecklund is the Herbert S. Autrey Chair, Professor of Sociology and Director of The Religion and Public Life Program at Rice University, where she is also a Rice Scholar at the Baker Institute for Public Policy. Elaine's research investigates how individuals develop cognitive schema that are at odds with institutions that constrain them. She is the author of two books with Oxford University Press and over thirty research articles. Her latest book is *Science vs. Religion: What Scientists Really Think* and she is currently writing a book entitled *Beyond the Ideal Scientist* in which she examines how women and men at elite research universities negotiate careers as academic scientists alongside family life.

Claudia Geyer is senior researcher at Professor Doris Lewalter's Chair for Upper Secondary School Education at the Technical University Munich. After working as junior researcher at the University of Federal Armed Forces in Munich and RWTH Aachen, she completed her doctorate at Technical University Munich. Her research involves analysing how to support interest in science through museum visits and how to foster public understanding of science and research by different kinds of learning environments.

Margaret Glass is Program Manager at the Association of Science-Technology Centres (ASTC) in Washington DC. Her work in the science museum sector has involved projects on astronomy and nanotechnology and she currently directs two programmes: Learning Labs in Libraries and Museums for the MacArthur Foundation and Institute of Museum and Library Services; and Intersections for the National Science Foundation Informal Science project. Margaret has published in a range of journals and edited collections, including the *Journal of Field Archaeology*, *PLoS Biology*, *COPUS Clarion*, and *Dimensions*.

Thomas Guterbock is Director of the Centre for Survey Research (CSR), Professor of Sociology and Research Professor of Public Health Sciences at the University of Virginia. He has been involved in over 400 funded survey projects, covering telephone, mail-out, in-person, and Internet modes as well as multi-mode projects. He is a frequent presenter of research papers at the annual meetings of the American Association for Public Opinion Research and has served on the AAPOR Executive Council as the elected Standards Chair.

Ines Hülsmann is currently undertaking a PhD on the history of research promotion by German foundations. She works for the German Literature Archive and Coordination Centre for Scientific University Collections, where she explores the structures enabling humanities research. Her research interests include the sociology of knowledge, the history, philosophy and sociology of science, and the communication of scientific knowledge.

Jennifer Jiminez is a theatre artist, workshop facilitator and Co-Executive Director of Aiding Dramatic Change in Development (ADCID). She is also an Artist Educator for the Ontario Arts Council. Her research interests include the active participation of design elements in devised creation and integrating the transformative aspects of community arts with the aesthetics of collaborative performance. She is currently working to push interdisciplinary boundaries with performance, electronic and digital media and assistive technologies for the physically disabled through ADCID's project, Exploring Possibilities in the World of Alternative Augmentative Communication.

Anthony E. Kelly is Professor of Educational Psychology, College of Education and Human Development, George Mason University, Fairfax, VA. He has served with the National Science Foundation and the US Department of Education, and is Principal Investigator on a National Science Foundation-funded Science Across Virtual Institutions (SAVI) project with the University of Helsinki. His research interests include research methods and modelling design-based research approaches, as well as modelling in online learning.

Doris Lewalter is Professor for Upper Secondary School Education at the Technical University Munich. Her research involves analysing the conditions, processes and results of teaching and learning inside and outside the school environment from a motivational perspective. She also examines how out-of-school learning environments such as museums, science centres and student labs can support learning and motivation at secondary schools.

Bruce Lewenstein is Professor of Science Communication in the Departments of Communication and of Science & Technology Studies at Cornell University, Ithaca, New York. He works primarily on the history of public communication of science, with excursions into other areas of science communication (such as informal science education). He has also been very active in international activities that contribute to education and research on public communication of science and technology, especially in the developing world. Bruce has been editor of *Public Understanding of Science* and co-chair of a United States National Research Council study, *Learning Science in Informal Environments: People, Places, and Pursuits.* He is co-author of *The Establishment of American Science: 150 Years of the AAAS*, editor of *When Science Meets the Public*, and co-editor of *Creating Connections: Museums and the Public Understanding of Research.*

Lorraine Locke is Senior Lecturer and Course Director in Learning and Teaching at the London School of Hygiene and Tropical Medicine. She has degrees in Sociology and Social Research Methods and is currently undertaking an ESRC funded PhD at the University of Surrey using discourse analytic methods to explore the interactional construction of dreams. Her most recent publications are on academic literacy and research methods.

Simon Locke is Senior Lecturer in Sociology at Kingston University in South-West London. He has been involved in the field of public understanding of science since the late 1980s and styles his research as concerned with public meanings of science. He has published widely in refereed journals, including *Public Understanding of Science*, *Science, Technology, & Human Values*, *Sociological Review* and the *Journal of Contemporary Religion*. His most recent book is *Re-crafting Rationalization: Enchanted Science and Mundane Mysteries* (Ashgate 2011) and he is currently writing a book on comics (at last!).

Katrin Neubauer is currently undertaking a PhD at the Technical University Munich, where she is a research assistant at the faculty TUM School of Education, Department for Upper Secondary School Education. Her research involves analysing the conditions and the development of motivation, interest and learning processes in museums and other (in)formal learning environments, as well as PUS/PUR (public understanding of science/ research) and the evaluation of museums and exhibitions. She has published in the journals *Computers & Education*, and *Psychologie in Erziehung und Unterricht*.

Ana-Maria Raus is a technology and digital media researcher. Her recent research interests range from science communication to technology analysis, interactive media and media studies paired with science and technology studies. As an academic visitor at the Institute for Science, Innovation and Society (InSIS), Saïd Business School, Oxford University, she investigated practices of digital mapping of social controversies and the use of digital tools in social sciences. She has also conducted research on interactivity and enhanced publications at the Royal Netherlands Academy of Arts and Sciences in Amsterdam, with the eHumanities group.

Katherine Roach is a writer and independent researcher. She studied zoology at University College London, before developing a career in science and natural history documentary-making. She has a PhD in Science and Technology Studies from Nottingham University and is interested in how the cultural assimilation of science influences the kinds of narratives that we tell in fiction, faction and "fact". She writes popular material, which has been published in a variety of outlets including BBC Wildlife, BBC On-line, Channel4.com, Channel 5 and *New Scientist*, and she is currently working on a book on the detective as a cultural hero.

Stephen Sillett is co-executive director of Aiding Dramatic Change in Development (ADCID), an organisation that researches, facilitates and directs dialogue, drama and art processes for community development. He is a member of the International Association of Facilitators and directs InFusion Labs where theatre artists, therapists, scientists and social practitioners explore spatial approaches to exploration and discovery. He is exploring approaches that engage community members in conversations consciously orientated to maturing visions of the future and is currently working with McMaster University Department of Social Sciences on a two-year project *Mapping the Self in Public Space: People with Intellectual Disabilities' Everyday Use of City Space*. His research interests include the facilitation of non-verbal and spatial meaning-making practices within group workshops and the creation of interactive performance.

Jeff Teare is Co-director of Theatrescience and ex-Associate Director of Theatre Royal, Stratford East and the Young Vic. He has been involved in over 100 professional productions ranging from pantomime to Pinter and is also an actor, writer, musician, lighting designer, teacher and video maker.

Chris Toumey is a cultural anthropologist who works in the anthropology of science and is a Research Professor in the University of South Carolina NanoCenter. He is interested in the cultural dynamics of public scientific controversies and is currently examining religious reactions to synthetic biology. Since 2003, he has focused on societal and cultural issues in nanotechnology including: religious reactions to nanotech; contested histories of nanotech; finding ways for non-experts to have active and constructive roles in policy discussions; and problems in the relation between a nanoscale object and an image of the object. He and has published on a range of topics, including creationism, cold fusion and fluoridation, and his commentaries on nanotech appear four times a year in *Nature Nanotechnology*.